Contents

Preface		*xi*
Acknowledgements		*xiii*
1	**INTRODUCTION**	**1**
1.1	The development process and recreation provision	1
1.2	Sport, leisure, recreation and recreation facilities	2
2	**THE ORIGINS AND EVOLUTION OF RECREATION PROVISION**	**7**
2.1	Introduction	7
2.2	The impact of industrial transformation and urbanisation	8
2.3	Early state intervention	11
2.4	Recreation as a merit good	13
2.5	Recreation provision as social palliative	16
2.6	The growth of commercial provision	18
2.7	The compulsory competitive tendering of local government services	20
2.8	Conclusions	21
3	**PROPERTY, PROPERTY MARKETS AND DEVELOPMENT**	**23**
3.1	Introduction	23
3.2	The allocation and exchange of property rights	34
3.3	The development process	41
4	**THE DEVELOPMENT PROCESS AND THE CONSEQUENT DEMAND FOR LAND**	**49**
4.1	Demand, need and the development process	49
4.2	The development intention	50
4.3	Feasibility	64
4.4	Financial viability	76

4.5	Construction	105
4.6	Monitoring and evaluation	107

5	STATE REGULATION OF DEVELOPMENT: THE STATUTORY PLANNING SYSTEM	116
5.1	Introduction	116
5.2	Development plans and development control	118
5.3	Evolution of the town and country planning system	120
5.4	Conclusion: the place of recreation provision in the statutory planning system	135

6	RURAL RECREATION AND FARM DIVERSIFICATION: CONTROL OF DEVELOPMENT IN THE COUNTRYSIDE	138
6.1	Introduction	138
6.2	Outdoor recreation and the demand for access	139
6.3	The supply of farm-based recreation opportunities	147
6.4	Water-based recreation	151
6.5	Conclusions	159

7	URBAN LEISURE PROVISION AND THE ROLE OF THE PUBLIC SECTOR	164
7.1	Introduction	164
7.2	The economic rationale for public leisure provision	165
7.3	The allocative efficiency of public leisure policy	172
7.4	The role of leisure facilities in the fulfilment of the Sport for All policy	175
7.5	The distributional equity of state intervention in the leisure market	176
7.6	The compulsory competitive tendering of local authority leisure facilities	179

8	THE ENVIRONMENTAL ASSESSMENT OF RECREATION DEVELOPMENT	182
8.1	Introduction	182
8.2	Environmental assessment in practice	184
8.3	The environmental impact of recreation development	187
8.4	The impact of statutory environmental assessment on recreation development in Britain	195
8.5	Conclusions	196

Recreation Planning
and Development

Greg Lloyd

Macmillan Building and Surveying Series

Series Editor: Ivor H. Seeley

Emeritus Professor, Nottingham Polytechnic

Advanced Building Measurement, second edition, Ivor H. Seeley
Advanced Valuation Diane Butler and David Richmond
An Introduction to Building Services Christopher A. Howard
Applied Valuation Diane Butler
Asset Valuation Michael Rayner
Building Economics, third edition Ivor H. Seeley
Building Maintenance, second edition Ivor H. Seeley
Building Procurement Alan Turner
Building Quantities Explained, fourth edition Ivor H. Seeley
Building Surveys, Reports and Dilapidations Ivor H. Seeley
Building Technology, fourth edition Ivor H. Seeley
Civil Engineering Contract Administration and Control Ivor H. Seeley
Civil Engineering Quantities, fourth edition Ivor H. Seeley
Civil Engineering Specification, second edition Ivor H. Seeley
Computers and Quantity Surveyors A. J. Smith
Contract Planning and Contract Procedures, third edition B. Cooke
Contract Planning Case Studies B. Cooke
Design–Build Explained D. E. L. Janssens
Development Site Evaluation N. P. Taylor
Environmental Science in Building, third edition R. McMullan
Housing Associations Helen Cope
Housing Management – Changing Practice Edited by Christine Davies
Introduction to Valuation D. Richmond
Marketing and Property People Owen Bevan
Principles of Property Investment and Pricing W. D. Fraser
Property Valuation Techniques David Isaac and Terry Steley
Public Works Engineering Ivor H. Seeley
Quality Assurance in Building Alan Griffith
Quantity Surveying Practice Ivor H. Seeley
Recreation Planning and Development Neil Ravenscroft
Small Building Works Management Alan Griffith
Structural Detailing, second edition P. Newton
Urban Land Economics and Public Policy, fourth edition P. N. Balchin, J. L. Kieve and G. H. Bull
Urban Renewal – Theory and Practice Chris Couch
1980 JCT Standard Form of Building Contract, second edition R. F. Fellows

Series Standing Order
If you would like to receive future titles in this series as they are published, you can make use of our standing order facility. To place a standing order please contact your bookseller or, in case of difficulty, write to us at the address below with your name and address and the name of the series. Please state with which title you wish to begin your standing order. (If you live outside the United Kingdom we may not have the rights for your area, in which case we will forward your order to the publisher concerned.)

Customer Services Department, Macmillan Distribution Ltd
Houndmills, Basingstoke, Hampshire, RG21 2XS, England.

Recreation Planning and Development

Neil Ravenscroft

Department of Land Management
and Development
University of Reading

MACMILLAN

First published 1992 by
THE MACMILLAN PRESS LTD
Houndmills, Basingstoke, Hampshire RG21 2XS
and London
Companies and representatives
throughout the world

ISBN 0–333–51880–2 hardcover
ISBN 0–333–51881–0 paperback

A catalogue record for this book is available from the British Library

Typeset by Ponting–Green Publishing Services, Sunninghill, Berks

Printed in Hong Kong

To John
From Neil and Jo

9   THE INTERACTION BETWEEN RECREATION
     DEVELOPMENT PROPOSALS AND THE
     PLANNING SYSTEM   199

9.1 Introduction   199
9.2 Thorpe Park   202
9.3 Somerford Keynes Holiday Village   207
9.4 Guildford Leisure Centre   211
9.5 Royalty and Empire   215
9.6 Conclusions   219

10   CONCLUSION: RECREATION PLANNING AND
     THE DEVELOPMENT PROCESS   223

References   *227*
Index   *245*

Preface

It is only comparatively recently that the discipline of recreation studies has existed in its own right. Prior to this practitioners, students and scholars were forced to use information and experience from other disciplines, such as geography, and other professions, such as physical education and rural surveying. Over the past ten to fifteen years this situation has vastly improved. On the one hand there is now a range of texts on recreation management in both urban and rural contexts, whilst on the other the expansion of education and training courses suitable for those wishing to join the Institute of Leisure and Amenity Management has greatly widened access to, and preparation for, a career in recreation management.

However, in a discipline as new as recreation there are bound to be gaps, both in terms of professional coverage and educational and training materials. One such area is addressed in this book, with its focus on the planning and development of recreation facilities. Whilst the nature of the property development process is well-known to most Chartered Surveyors and surveying students, the nature of recreation property is virtually unknown to most recreation practitioners and students.

These are significant gaps, given the central importance of planning and developing land for recreation, especially when it is recognised that the development of recreation property is, in several respects, unique. Most fundamentally, it differs from many other types of property development by not being speculative. Rather than a developer determining the physical location and characteristics of a building, as is the case in most house, office and industrial development, recreation buildings are usually commissioned by the occupier. This has led, not unnaturally, to the view within the property world that the value of recreation buildings is highly dependent upon their management; a situation not generally found in other types of property. Quite apart from the difficulties that this creates in obtaining funding for the development of such buildings, there is the associated difficulty of future inflexibility of use, whether notional or real.

Added to the uniqueness of recreation property is the place that such property has played in the development of national land policy in Britain. In particular, the provision of sports facilities, usually in the public sector, has frequently been linked to the pursuits of other goals, such as social stability or the amelioration of the effects of long-term unemployment. Equally, policies affecting recreation and tourism development in the private sector have mostly been concerned with regional development, employment generation and contribution to the balance of payments. Finally, recreation provision has also been an important element of rural policy, but largely as a means of retaining the *status quo* of existing land ownership and land use patterns.

These characteristics of the planning and development of recreation property make it an extremely important area of study for practitioners and students alike. In this text, developed largely from undergraduate, post-graduate and professional courses in recreational land management and development at the University of Reading, I have attempted to draw the various strands of recreation, planning and development together in a way that is accessible to a wide range of disciplines and professions, including surveying, recreation, planning, geography and government, in the hope that it can contribute to the growing debate on the place of recreation in people's lives.

Neil Ravenscroft
Reading, December 1991

Acknowledgements

As with any such project, numerous people have been consulted and have given freely of their time. Without diminishing the contribution of any of them, some do deserve a special mention. Of particular note has been the input to the work on planning provided by Dai Edwards; without his timely contribution of material and constructive comment the book would surely not have been completed in its present form. A special note of thanks is also due to Peter Byrne with whom I worked on the research into farm diversification which is referred to in Chapter Six. My thanks are also due to those who agreed to the inclusion of the case studies and provided much valuable information about them: Derek Oliver of Leisure Sport Ltd, for both agreeing to the inclusion of the Thorpe Park case study and for providing valuable information for, and comment about, the relevant section; Huw Bannister of Guildford Borough Council, for agreeing to the inclusion of, and providing comment about, the Guildford Leisure Centre case study; Bill Andrewes of the Granada Group plc, for agreeing to the inclusion of the Somerford Keynes Holiday Village case study; Jim Gosling for providing information and comment about the Somerford Keynes Holiday Village and the Guildford Leisure Centre; and Geoff Messenger, for information and comment about Royalty and Empire. Lastly, special thanks are due to Sarah Colledge for processing the tables and figures as well as deciphering a range of computer disks in order to produce a final copy of the work.

This text leads on from the book *Outdoor Recreation and the Urban Environment* written by Professor Ivor Seeley, the series editor, as long ago as 1973.

Acknowledgements

1 Introduction

1.1 The development process and recreation provision

In a country such as Britain, where recreation facilities, ranging from municipal parks to leisure centres and commercial theme parks, are now relatively commonplace, it is important to recognise just how recent a phenomenon this is. With the exception of urban parks and some swimming baths and playing fields, little of this provision existed thirty years ago. In those thirty years, the number of publicly provided sports centres and swimming pools has grown to around 1000 nationwide, whilst expansion of the number and range of facilities in the private sector has been much more rapid and widespread.

By the mid- to late-1980s, however, it had become evident that recreation provision had reached something of a milestone, or crossroads. On the one hand, the 'leisure society' upon which much of the private sector development was predicated has never truly materialised, although the short-lived consumer boom of the mid-1980s may have given rise to some renewed expectations. Similarly, the expectation, on the part of the public sector, that leisure provision would become an accepted and fundamental part of the Welfare State, so enhancing the status of leisure provision, has never been realised, although even if it had the very future of the Welfare State is now in the balance. Of equal significance in this debate has been the volatility of the tourist industry, responding as it does to worldwide social and political events. There is also a growing concern over the potential environmental impact of leisure, recreation and especially tourist activities, both domestically in areas such as the Lake District and the Peak District, and internationally in the degradation of areas such as the Mediterranean and its surrounding coastline.

The recognition of these problems has given rise to a major debate over the future of leisure services and their provision. Current conferences and congresses, including the World Leisure and Recreation Association's 1991 congress in Sydney, Australia; the European Leisure and Recreation Association's 1992 congress in Bilbao, Spain and the Leisure Studies Association's 1993 international conference in Loughborough, England, have all taken the theme of providing for leisure and recreation in a changing world. Similarly, academics and practitioners in the United Kingdom have been undertaking a wide range of studies on the future of leisure provision, including examinations of resource implications (Patmore 1983 and Pigram 1983), economics (Gratton and Taylor 1985 and 1988) and management (Torkildsen 1986 and Henry 1990), as well as a host of other related issues. Much information is, therefore, now available on many aspects of recreation provision, whilst the management of facilities is fast becoming recognised as a serious profession as well as an established academic discipline.

An area that has yet to receive attention is, however, the process that determines what facilities are actually developed, as well as where, when, how and by whom. Whilst apparently fundamental to the whole process of providing facilities, the notable absence of this type of work to-date is probably based on the misconception that an economic analysis can answer all the questions related to demand, need and location, whilst the construction industry can be called upon to turn the wishes of the recreation providers into reality.

Although partly the case, this view takes little account of the role or importance of the development process in determining the final shape, form and location of individual leisure facilities. Whilst this omission might at first glance appear rather technical and mundane, issues such as the ownership and control of resources, the availability of finance and the influence of the statutory planning system indicate just how central the development process is to leisure provision in both the public and private sectors of the economy. Indeed, a failure to recognise the importance of these issues may well have contributed significantly to the present problems being experienced in public and private sector alike. This is particularly so in the way in which recreation provision has constantly struggled for recognition in the statutory planning process, as well as the continuing indifference shown by City financiers to underwriting leisure developments.

In an attempt to rectify some of these omissions, this book seeks to examine the nature of the property development process as it relates to both public and private sector recreation provision. In doing so, the book will examine the nature of property, property ownership and the consequent property power, together with the constituent elements of the development process and the consequent demand for land. A detailed examination of the statutory planning process will also be undertaken in an attempt to analyse how and why the state seeks to regulate such development and what effect this has had on recreation provision in both urban and rural areas. Finally, the analysis will be applied to four contrasting leisure developments as a means of illustrating the stages of the development process, the relationship between developers and the state and the ways in which the process of development has altered over the past twenty years.

1.2 Sport, leisure, recreation and recreation facilities

Many authors have attempted definitions of sport, recreation and leisure, including Seeley (1973), Parker (1976), Roberts (1978) and Pigram (1983).

However, as Neulinger (1982) and Gratton and Taylor (1985) state, there are many pitfalls involved in such definitions, particularly those relating to the problem of defining a concept that is personal to each and every individual. Not withstanding this caution, it is important that a working definition of these concepts is established at the start of this book. This is not done in a misguided attempt to add to or refine the existing literature, but as a means of establishing precisely what is and, consequently, what is not to be discussed in the forthcoming chapters.

There appears to be some consensus in the literature as to working definitions of the terms, based largely on the work of Parker (1976). The definition of leisure has three main elements: leisure as time; leisure as activity; and leisure as a combination of both. In terms of time, leisure can be seen as a residual after 'necessary' activities have taken place. These 'necessaries' include sleeping, eating for survival, personal hygiene and work, if any. In terms of activity, conversely, leisure 'encompasses activities which are characterised by a feeling of comparative freedom' (Parker 1976, 12). The third element of the definition of leisure seeks to combine both time and activity. Thus, there is firstly a residual time element which is then qualified by a normative statement of what leisure ought to be (Parker 1976, gives some examples). The length of time available governs the amount of activity undertaken; the activity itself is dependent upon both the individual and upon societal influences, including such things as motivation, education and expectation.

Whilst there is general agreement in the literature that 'recreation' represents a part of 'leisure', there are differences of opinion as to which part. For example, Pigram states that 'recreation is considered to be activity voluntarily undertaken, primarily for pleasure and satisfaction, during leisure time' (1983, 3). In contrast, McCormack states that:

> Recreation is a system of social control, and like all systems of social control, it is to some degree manipulative, coercive and indoctrinating. Leisure is not.
> (1971, 171).

Without agreeing with either of these statements, Arnold (1985) suggests that the meaning of 'recreation' is conditional on a given value system. If 'recreation' is taken to represent the re-creation or revitalisation of the individual after participation in certain activities, a value or worth is attributed to 'recreation', in the sense that creativity is good and non-creativity is bad. There is, however, little discussion of how this value has become attributed to recreation (Arnold 1985). Consequently, Pigram's (1983) definition is the more universally accepted, with authors such as Patmore (1983) describing recreation as a specific activity undertaken in leisure time and Chubb and Chubb (1981) considering recreation to be 'any type of conscious enjoyment'. In adopting these definitions for the purpose of this book, any activity or inactivity undertaken for enjoyment could legitimately be considered to be 'recreation'.

In common parlance, 'sport' is normally considered to be synonymous with competitive recreation activities, whether active or passive. Whilst this certainly covers many of the attributes associated with sport, a more formalised and wider definition is contained in the European Sport For All Charter:

> A great range and variety of activities is covered by the term sport; they subdivide into four broad categories:
>
> 1 Competitive games and sports which are characterised by the acceptance of rules and response to opposing challenge.
> 2 Outdoor pursuits in which participants seek to negotiate some particular 'terrain' (signifying in this context an area of open country, forest, mountain, stretch of water or sky). The challenges derive from the manner of the negotiation adopted and are modified by the particular terrain selected and the conditions of wind and weather prevailing.
> 3 Aesthetic movement which includes activities in the performance of which the individual is not so much looking beyond himself and responding to man-made or natural challenges as looking inward and responding to the sensuous pleasure of patterned bodily movement, for example dance, figure-skating, forms of rhythmic gymnastics and recreational swimming.
> 4 Conditioning activity, i.e. forms of exercise or movement undertaken less for any immediate sense of kinaesthetic pleasure than from long-term effects the exercise might have ... improving or maintaining physical work capacity and rendering subsequently a feeling of general well-being. (Council of Europe 1976).

Thus the concepts of time and state of mind characterise all three definitions, from the broad aspects of leisure through to the more narrowly defined limits of sport. Only the definition of sport is based upon a categorisation of activity, although in practice many recreation pursuits may also involve an element of activity.

It is often thought that leisure represents an all-embracing concept, of which recreation and sport are smaller sub-sets. These definitions suggest that this is erroneous, since recreation need not be limited solely to leisure time, and sport can be undertaken as a profession rather than as a use of residual time. Equally, the boundaries between the terms are sometimes unclear, with inactive recreation being similar to leisure and active re-creation, such as walking, being similar to sport. For these reasons it is probably most helpful, in the context of this book at least, to think of leisure, recreation and sport representing a single continuum, progressing from leisure through recreation to sport, with the changes from one category to another being gradual and ill-defined.

From this discussion, a leisure, recreation or sports facility can be thought of as any medium offering the opportunity to participate in leisure, recreation or sport. Indeed, the *Oxford Dictionary* definition of 'facility' includes the phrase 'unimpeded opportunity'. Dower *et al.* (1981) note that resources must be utilised to provide a facility and that such facilities can be simply divided between those provided primarily for purposes other than leisure activity (such as woods, rivers and streets), and those provided primarily for leisure, recreation and sport (such as sports centres and swimming pools). In a similar vein, Clawson and Knetsch (1971) outlined three categories of recreation area: resource based; intermediate; and user based. The first of these refers to natural resources where leisure activities take place, such as woods, rivers and scenic areas; the last one refers to facilities specifically developed for recreation and leisure activities, such as the sports centres and swimming pools mentioned above; whilst the middle category relates to facilities between these extremes, such as country parks close to urban centres, where the environment has been modified, if not created. From these definitions it can be stated that leisure and recreation facilities include any resource used to facilitate the leisure-sport continuum in its broadest context.

Many authors have developed taxonomies of leisure facilities (Roberts 1979, Coopers & Lybrand Associates Ltd 1981, Dower *et al.* 1981 and Stabler 1982, for example), whilst others have developed taxonomies for other sorts of facility (Kouskoulas and Koehn 1974 and Meggitt 1980). However, in none of the cases have the taxonomies sought to cover the complete spectrum of leisure or recreation facilities, as Table 1.1 is designed to do. In presenting such a wide-ranging taxonomy the intention is not to suggest that all the facilities listed are necessarily created through the process of development, but to illustrate where such facilities fit within the complete spectrum of resources that can be defined as leisure or recreation facilities.

Table 1.1 A taxonomy of leisure facilities

Facilities not existing primarily for leisure	Resource-based facilities adapted for leisure	Built facilities adapted for leisure	Built facilities designed for 'passive' leisure	Built facilities designed for 'active' leisure
Agricultural land	Woodland parks	Historic houses	Museums	Marinas
Commercial woodland	Urban/rural parks	Ancient monuments	Galleries	Leisure/sports centres
	Golf courses	Redundant churches	Libraries	Dance halls
Watercourses	Beaches	Warehouses/ industrial buildings	Arts/community centres	Squash/tennis centres
Water masses	Cruising waterways			Gymnasia
Private dwellings	Public footpaths and bridleways		Cinemas	Swimming pools
Workplaces			Restaurants	Holiday camps
Streets/pavements	Canal towpaths		Hotels	Snooker halls
Moorland/ mountains	Watercourses			Sports stadia
	Water masses			Playgrounds
Reservoirs	Zoos			'All weather' sports pitches
	'Theme' parks			Sports clubhouses
	Open air museums			'Theme' parks
	Holiday camps			

It can be seen in Table 1.1 that only two of the five categories contain purpose-designed facilities, whilst two other categories concern resources and buildings adapted for leisure and recreation uses. It should also be noted that facilities with the same name or description can appear in different categories, depending largely upon whether they have been purpose-built or adapted.

In considering the nature of provision, it is clear that the fourth and fifth categories, relating to facilities designed for leisure uses, must have involved some physical development, usually in the form of building construction. What is less clear is the extent to which the first three categories may involve development. Column one (facilities not existing primarily for leisure) is most easily dealt with because none of the facilities listed are really 'leisure facilities', but facilities in which leisure or recreation may occur. However, they exist, or are primarily provided, for other reasons or purposes. The recreational use of facilities in column one, therefore, is not relevant to the book. Facilities appearing in columns two and three, however, may have originally been provided or developed for other purposes, but in being adapted to leisure or recreation usage will almost certainly have undergone some form of development, usually of a physical nature. In overall terms, therefore, the facilities described and considered in this book will fall primarily in columns four and five of Table 1.1, supplemented by those in columns two and three. Finally, the existence of the leisure-sport continuum indicates that there will be few readily defined 'leisure facilities' as opposed to 'recreation facilities', meaning that for the purpose of the book, facilities for leisure, recreation or sport will be treated as synonymous.

2 The Origins and Evolution of Recreation Provision

2.1 Introduction

It is difficult to determine when the 'provision' for recreation first emerged. What is certainly apparent is that from medieval times the concept of recreation has been considered an integral, if not distinct, part of life. Much early recreation activity was probably no more than an extension of the necessities of life, such as hunting and fishing. Indeed, the triumvirate of hunting, shooting and fishing remain thought of today as 'traditional' country sports. Gradually the nature of recreation became more formal, with the more powerful members of society beginning to organise recreation activities to suit themselves, often at the expense of the powerless. Thus the spontaneity of hunting by everyone was replaced by the development of 'facilities', such as the hunting forests reserved for the aristocracy and the ruling classes.

Many further recreation activities became popular in early civilizations, including horse racing, boxing, wrestling and archery (Chubb and Chubb 1981). It was probably at this time that people first became employed to provide recreation activities for others, although it was likely to be as a result of force rather than free will or the exploitation of a market opportunity.

Perhaps the greatest advance in recreation provision was during the Greek civilization. The ruling class recognised the need for a fit and strong army to defend its position. It therefore promoted the desirability of physical fitness, praised athletic ability and provided a wide range of sports facilities to ensure that people had the opportunity to participate. In addition, the Greeks developed the concept of the leisure ethic, whereby the intelligent use of free time was considered to be the principal aim in life. This led to the provision of parks, theatres and a wide range of recreation facilities in most urban areas.

In common with the Greeks, the Romans encouraged physical fitness for health and military reasons. Rather than concentrate on the leisure ethic, however, the Romans developed the practice of tourism, with travel to the spa centres, to resorts and cultural attractions and, in some cases, to foreign attractions such as the Pyramids. At the same time there emerged in Rome an urban middle class that could not afford to undertake tourist activities, but which none the less had considerable amounts of free time. This free time was spent, in the main, by attendance at parades, horse races, exhibitions and various gladiatorial games, most of which were organised by the state.

Following the fall of the Roman Empire there was a period of relative austerity under a predominantly Catholic church, where the excesses of the Roman lifestyle were seen as things to be overcome. However, this did not stop the participation of the ruling classes in hunting, banqueting feasts or theatrical entertainment. In contrast, this period saw the emergence of what later became known as the Protestant work ethic, where the nobility encouraged ordinary people to believe that hard work was beneficial and necessary for the soul.

Following the decline of the Catholic church there emerged two opposing trends in recreation. The first saw the reintroduction of recreation activities, such as an interest in the arts and theatre by the educated and well-to-do, as well as many less salubrious pastimes such as gambling, cock fighting and bear baiting by the lower classes. Opposed to this was a group of people who continued to practice the Catholic ethics, eventually leading to the Protestant reformation and work ethic. However, the ethic had little effect on the wealthy or the nobility, who continued to do much as they had done before. It was at this time that many formal gardens and parks were created.

However, up to the end of the Eighteenth century very little had actually changed over the preceding centuries:

> ... only the ruling classes were able to enjoy a wide range of opportunities. In spite of the earlier Greek and Roman development of public recreation, there were virtually no lands or programs specifically dedicated to this purpose in later periods. The common people had to be content with some social drinking, a little gambling, an occasional festival, and such other amusements as they could themselves devise. (Chubb and Chubb 1981, 20).

Even publicly owned parks were surrounded by iron railings, with a charge made for admission, so confining their use to the wealthy members of the community (Seeley 1973).

2.2 The impact of industrial transformation and urbanisation

By the early Nineteenth century the Protestant attitudes to work and recreation were becoming eroded by the emergence of a new middle class intent upon enjoying some of the privileges previously reserved for the wealthy. In addition to the popularity of theatre, dance and other traditional recreation pursuits, the industrial revolution produced a number of important developments in recreation provision. Not the least of these was that new technology made both domestic and foreign travel accessible to the middle classes, so setting off a new wave of tourism not seen since the end of the Roman Empire. Equally, the same organisation that had been applied to industry was used by the middle classes to 'pioneer' the development of organised sport. Thus rules and conventions were laid down for many of the existing sports, preserving them in the process for particular social and gender groups.

The extent to which these enthusiasts took over the organisation of sport was examined by McIntosh (1966) who noted that, with the possible exception of tennis, no sports were actually invented at this time; rather, they were refined to suit the needs of the middle classes and, in particular, the enthusiasts concerned. This led to the development of single-sport governing bodies with wholly separate finances and control. These bodies also became very wary about intervention by other groups and, most particularly, by government. The remnants of this attitude remain today, as witnessed by the continued independence of bodies such as the Central Council for Physical Recreation (CCPR) and the British Olympic Committee.

However, industrialisation was not only responsible for the organisation of sport. Its influence also extended, to some extent, to the very purpose and ethos of sport. Previously, sport and recreation had been seen as pastimes, to be indulged in when work was done. Now they were also seen as part of the 'new' culture, in helping mould character, condition behaviour and improve social accomplishment. This was particularly so in education, where sport was seen as an ideal method of developing character. Indeed, sport became taken so seriously at some schools that commentators such as McIntosh (1966) claimed that the schools of Rugby, Eton, Harrow, Charterhouse, Winchester, Westminster and Shrewsbury, together with the Universities at Oxford and Cambridge, were responsible for establishing international sports competition.

Within the schools, sport was used to produce the particular range of skills, such as self-confidence and loyalty, that the middle class felt necessary to protect their position. Thus sport became an integral part of the training of future public servants, industrialists and, of course, army officers. The link between sport and military power has remained an important concept since the days of the Roman Empire. Indeed, in the period 1830 to 1860, when most of the modern-day sporting associations were formed, the British Army managed to fight 11 wars.

Alongside militarism came an upsurge of nationalism, which was fostered through sport, particularly with regard to international competition. This led directly to the reintroduction of the modern Olympic games. As this situation implies, Britain was not alone in developing sport at this time. Germany had introduced gymnastics at school as part of the preparation for the 'war of liberation' against Napoleon. Similarly, gymnastics were compulsory in Russian schools by 1865 and the French were using sport as a means of regenerating the French nation. In Britain itself, the Education Act 1870 was explicit in its use of gymnastics as a means of increasing work, military power and health, as well as being a means of reducing the expense of the criminal machinery. Indeed, this *raison d'être* for sport has never really changed. It remains with us in the late Twentieth century through the aims of the Sport For All policy in Britain, the National Fitness Council in the USA and Canada's Task Force on Sport, for example.

The elitism enshrined in the organisation of sport and recreation remained through much of the Nineteenth century, with artisans excluded from much of it, even if they had the time and money necessary to participate. However, changes were at hand and the last quarter of the Nineteenth century witnessed a minor revolution in sport and recreation. By 1870 most working people in Britain got Saturday afternoon off work. This led to a great proliferation of recreation activities, clubs and societies, as well as interest in pursuits such as allotment gardening (Seeley 1973). Much of this activity was actually provided by the middle classes, for it was very quickly realised that it was imperative to channel the workers' energy into wholesome activities, such as sport, recreation and leisure activities. Thus, alongside the monotony of factory work came the chance for self-expression through sport and recreation. Yet in neither case were the middle class far away. By owning not only the factories (and therefore being responsible for the workers' livelihoods) but also the sports and recreation grounds, the middle class continued to dominate, even if on the surface there had been some democratisation. This was further enforced by the arrival of professionalism in sport at this time. Whilst it appeared to offer some escape for the workers, it too was developed and controlled by the middle class, almost as a means of reducing working class sport to no more than just another method of production. In this period, therefore, sport was used as a method of social control for the workers just as surely as it was used as a method of social conditioning for the middle classes.

Alongside the 'democratisation' of urban sport and recreation, an increasing number of people were seeking to escape from the urban areas during their free time. In Britain this led to the start of a long battle for a right of access to open country whilst elsewhere, and particularly in North America and Australia, rural areas were being set aside for conservation purposes as well as recreation opportunities. In Britain, the issue of access gained in importance by being seen as a battle between the landless workers and the landed aristocracy. In a succession of Bills put before Parliament the access lobby attempted to restrict landowners' rights of exclusion, on the grounds that:

> ... no owner of uncultivated mountain or moorland should be entitled to exclude any person from walking on such lands for the purposes of recreation or scientific or artistic study. (Stephenson 1989, 40).

During the last quarter of the Nineteenth century many people organised themselves into rambling clubs. These clubs were to be found all over England, with a proliferation in the large urban centres. By the turn of the century many of these clubs began to unite, to form groups such as the London Federation of Ramblers and the Manchester and District Federation of Ramblers. However, it was not until the third decade of the Twentieth century that the groups united to form the Ramblers' Association.

2.3 Early state intervention

The last quarter of the Nineteenth century witnessed the return of the state to the provision of facilities for sport and recreation. Not since the decline of the Roman Empire had the state really provided facilities for ordinary people, but the onset of industrialisation and the attendant change in people's lives produced a need for government to re-examine its role in society.

Initially the re-emergence of state involvement came about as a result of public health legislation, with local councils being given the ability to provide sports facilities such as playing fields and indoor pools. The playing fields were often acquired from, or the responsibility for maintenance taken over from, factory owners and philantropists. The pools, conversely, were mostly built by the local authorities themselves. Similar events occurred elsewhere, particularly in North America. Having already accepted that physical education should be an integral part of schooling, cities such as New York, Boston and Chicago developed facilities such as playgrounds, sports grounds and small country parks. However, state intervention also came about as a result of the poor physical condition of many of the potential recruits to the Army at the time of the Boer War. The subsequent establishment of the Boy Scouts and Girl Guides in the first decade of the Twentieth century, with their emphasis on outdoor physical pursuits, confirmed the Nation's, if not directly the state's, concern with physical fitness (Seeley 1973).

Alongside provision, the state also began to accept the value of conserving the countryside, at least as far as responding to the wishes of some voluntary groups. In Britain this led to the formation of the National Trust in 1895 by an Act of Parliament, with the intention of restricting the loss of urban commons as well as acquiring beautiful countryside to serve as an 'open air sitting room for the poor'. In contrast, many States within the USA passed legislation facilitating the purchase of countryside and the establishment of parks and recreation grounds by the state itself.

The first quarter of the Twentieth century saw a significant improvement in people's living environment, with reductions in the working week as well as better education, health care and housing. When combined with improved transport, particularly in the form of rail and bus travel, as well as the bicycle, the demand for recreation and sports activities increased significantly. Provision continued, however, to reflect the haphazard nature of its introduction. Thus, whilst local councils would provide land or other facilities, it was still left largely to voluntary clubs to organise activities (Bacon 1980, Blackie *et al.* 1979 and Travis 1979 give more details).

Inevitably this lack of co-ordination tended to mean that recreation provision was accorded a low priority when the state was planning its future use of resources. This led, in 1935, to the formation of the Central Council of Recreative Physical Training (becoming the Central Council for Physical

Recreation in 1944) as the 'umbrella' body for all those involved in the administration of sport. This coincided with the formation of the Ramblers' Association, also in 1935, and signalled much increased pressure on the state to provide for sport and recreation.

This pressure culminated, in 1937, with the Physical Training and Recreation Act and, in 1939, with the Access to Mountains Act. The first of these statutes improved the power of local authorities to give financial aid for recreation facilities at schools and within the community. However, the effectiveness of this Act was never tested due to the start of the Second World War just two years later.

Passed on the eve of war, the second statute, in name at least, represented a successful end to over half a century of campaigning for free access to mountains and moorlands. However, what started at the Bill stage as an access charter became, upon enactment, a landowners' protection charter (Stephenson 1989). By empowering landowners and ramblers to agree access areas, but at the price of those entering land not subject to an access agreement being guilty of the offence of trespass, the Act gave new powers to landowners not only to exclude the public, but also to forcibly remove offending trespassers from their land. Once again the Second World War intervened, allowing time for both the ramblers and landowners to review the likely effect of the legislation and, on the ramblers' part at least, what they would do to improve public access to private land.

By 1942 Britain was beginning to make plans for restructuring once peace had been achieved. Reports by various people including Dower, Scott and Hobhouse considered issues such as the future control of development (culminating in the Town and Country Planning Act 1947) and the creation of national parks and the securing of public access to the countryside (culminating in the National Parks and Access to the Countryside Act 1949). However, as far as sport was concerned, perhaps the most fundamental change to have occurred at that time was the provision for compulsory physical training in all schools, passed in the Education Act 1944. Thus, after being part of public school curricula for the past 100 years, compulsory physical training in schools became the first major incidence of state intervention in the provision of facilities for sport and recreation in Britain.

That the state did not extend its intervention to granting free access to mountain and moorland under the National Parks and Access to the Countryside Act 1949 is indicative of the way in which public policy was to develop over the next 30 years. Whilst encouragement to provide facilities and participate in urban sport was part of state policy, so was the restrictive attitude towards recreation in the countryside and the protection of private property rights.

Following the aftermath of the War and the creation of the Welfare State there followed a quiet period for public recreation provision. However, experience of travel during military service, or factory work as part of the war effort, had either expanded people's horizons or had convinced them of the need to do so. Thus, as resources became available the demand for travel began to increase significantly, heralding what Chubb and Chubb (1981) see as the 'mechanisation of recreation'. As more people could afford to own a car, or could afford coach fares, visits to the coast, national parks, historic sites and the countryside increased rapidly. This soon led to serious congestion at many popular locations, as well as forcing a reconsideration of the place of sport and recreation in people's lives.

2.4 Recreation as a merit good

The impetus for a review of sport and recreation came not from government, but from the voluntary sector, in the guise of the Central Council for Physical Recreation, who commissioned the Wolfenden Committee in 1959. Its report (Wolfenden Committee 1960) indicated that the existing structure of recreation provision and sports administration was sound, but that a new, state-appointed, advisory planning service was required. These findings were duly endorsed by all the political parties and, in 1965, the Advisory Sports Council was set up, largely under the control of the Central Council for Physical Recreation. So, whilst the state appeared to have been involved in the review and restructuring of recreation provision, it had in reality done little more than confirm and strengthen the role of the voluntary sector, which was itself still safely controlled by those same middle classes that had first organised sport in Britain.

Elsewhere, similar reviews were being undertaken. However, in the USA for example, the federal government was responsible for a 28-volume report from the Outdoor Recreation Resources Review Committee (1962), and for implementing its suggestion for a federal Bureau of Outdoor Recreation to co-ordinate recreation provision at all levels of government. Similarly, the French government undertook a review to establish how to provide for 'leisure for all'.

Towards the end of the 1960s the Countryside Commission came into being as a replacement for the old National Parks Commission. This appeared to signal a shift towards the provision of improved recreation facilities in the countryside, through a new system of grant aid and advice to local authorities and private landowners, contained in the Countryside Act 1968. Indeed, this did initially appear to be so, with generous grants helping local authorities establish many new country parks and picnic sites set aside for informal recreation. However, little was done to improve the wider issues of access to the countryside or to set aside more areas for the nation.

The effect of this reorganisation of the provision of both urban and rural recreation is difficult to judge. That the 1960s saw the start of a major period of facility development is without doubt, but the extent to which this reflected the reorganisation or was largely independent of it is questionable. Certainly, the first indoor sports centres, such as that at Harlow, were already open by the time the Advisory Sports Council was formed, and some eight years before it produced guidelines for the number and type of facilities required by the nation (Sports Council 1972). Furthermore, since the Advisory Sports Council was still ostensibly controlled by the voluntary sector it tended to do little more than reflect their existing policies and ideologies. Thus space and provision standards suggested by bodies such as the National Playing Fields Association were accepted as objective standards, without any question as to their validity. Equally, the impending reorganisation of local government at the start of the 1970s precipitated a large volume of spending by these authorities prior to their demise, with much of it being invested in the provision of new recreation and leisure facilities. As Bacon (1980) recounts, in 1970 there were just 27 sports centres in England, but by March 1974 the number had grown to 167 already open with another 273 either planned or under construction. Included amongst these were an increasing number of facilities built on school premises but open to the whole community (Seeley 1973). However, gaining acceptance for this type of use proved slow, with objections over the need for increased supervision, greater wear and, in the case of sports pitches, that they were already used by other organisations, such as youth clubs (Seeley 1973).

The 1960's also saw a great increase in tourism, both domestically within Britain and from foreign countries. In order to make the most of the opportunities offered by this new industry, Parliament passed the Development of Tourism Act 1969 and with it created the English Tourist Board. In common with the Countryside Commission and, from 1972, a new independent Sports Council, the English Tourist Board saw a large part of its job as administering grant aid to encourage the development of new and improved facilities. However, whilst the Countryside Commission and Sports Council, with responsibility to the Department of the Environment, were part of an administrative chaos that had grown up around the provision of facilities for sport and recreation in the public and private sectors of the economy, the English Tourist Board immediately took up a central place in the economy, becoming part of the Board of Trade's machinery for regional economic development.

By 1973, in spite of the existence of the Regional Councils for Sport and Recreation, originally set up alongside the Sports Council to provide a regional forum for co-ordinating the provision of recreation facilities, there was a growing awareness that, whilst an increasing number of local authorities were spending a great deal of ratepayers' money on recreation provision, the administrative framework within which they were operating was far from explicit. Indeed, in a House of Lords Select Committee report (House of Lords 1973) it was noted that since sport and leisure were largely undefined in governmental terms there could be no coherent means of administration. This was exacerbated by the aforementioned reorganisation of local government in 1974, where the responsibility for providing recreation facilities was not given to either of the new tiers of local government. This inevitably led to both county and district councils providing recreation facilities, sometimes resulting in a duplication of provision. Thus recreation provision had apparently become a politically peripheral issue, with no overall responsibility at either central or local levels of government, regardless of the fact that most government departments had some obligations or responsibilities for its provision. Yet, both the House of Lords and, in a separate paper two years later, the government (Department of the Environment 1975), had accepted that sport should rightfully be seen as an integral part of the Welfare State, despite their continued failure to reserve adequate funding for sport:

> Sport and recreation are not luxuries, they are an essential part of life. In industrialised societies with all the pressures of living in towns, they are more important than ever in helping to preserve physical capacity and self-expression. Central and local government should recognise it as their duty to support the development and promotion of sport. It should not be necessary for sport to have to press for a gambling levy or to rely upon a system of national lotteries. (Roger Bannister, Chair of the Sports Council, quoted in Sports Council 1974, 3).

The confusion over responsibility for provision did not stop the government from detailing in its White Paper (Department of the Environment 1975) the important issues to be addressed in the provision of sports facilities by the public sector. These included adequate provision in inner city areas, provision for school leavers, disabled people and for gifted athletes. Against these issues and the background of confusion, local authorities were swift to recognise the vote-winning potential of developing large sports facilities, especially when those facilities could be linked to civic pride.

The provision of sports facilities continued apace during this period, but with no thought for co-ordination. This inevitably led to an over-supply of facilities in some areas, where neighbouring councils all provided similar facilities, and an under-supply in others, where no provision had been made. In recognising this chaos, Dennis Howell, then Minister for Sport, took steps to strengthen the role of the Regional Councils for Sport and Recreation by charging them with a duty to develop regional plans for sport and recreation which would form an important input to the County structure planning process (an example of this can be found in Northern Council for Sport and Recreation 1982).

Faced with this important task, the Regional Councils were quick to start producing technical documents on the current standards of provision and the consequent shortfalls. However, the actual process of agreeing on an overall plan proved to be much more contentious. This inevitably led to delays which meant that the County structure planning process went ahead without any formal guidance from the Regional Councils for Sport and Recreation. Denied the opportunity to influence the structure plans, the Regional Councils lost their impetus, so that by the early 1980s no regional plans had yet been produced.

With the benefit of hindsight it is possible to see that this chance to develop regional plans could have provided the mechanism to ensure that recreation provision formally became part of the Welfare State. As it was, the challenge was not accepted and, soon afterwards, the opportunity was lost as the country slid into recession, and the Labour administration was replaced by a new Conservative regime intent on reducing the role of government in the economy.

2.5 Recreation provision as social palliative

After the Conservative election victory in 1979 recreation provision could have been a very 'soft' target for government cuts, since local authorities were not mandated to provide particular levels of service, as they were in most other areas of their work. However, as the jobless numbers began to rise at the start of the 1980s, the Sports Council was quick to point out the potential role of sport in ameliorating the boredom of the unemployed (Sports Council 1982). Equally, the English Tourist Board was quick to seize upon the potential of recreation and tourism to create employment in a new expanding industry; a recognition that later caused it to be moved from the control of the Department of Trade and Industry to the Department of Employment.

Initially, it appeared that the argument proffered by the English Tourist Board would prove most persuasive to a government committed to a market economy. Indeed, the idea of recreation becoming part of a new service industry rather than a dubious 'add-on' to the Welfare State held much appeal and has, subsequently, become an important part of government policy. However, the role of government in recreation provision was brought to the fore again in 1981 with the civil unrest in London and Liverpool. The subsequent investigation into the trouble (Scarman 1981) indicated, as had earlier been evidenced in the first stages of the development of New Towns in the 1950s, that much of the problem centred around young jobless people with too little to keep them occupied. As a result, the report recommended the provision of neighbourhood sports and recreation facilities as a means of providing things to do. This resulted in Urban Aid grants for developing facilities (Carrington and Leaman 1983) and an ambitious new development programme by the Sports Council to provide small standardised 'value-for-money' sports centres for individual communities (Sports Council 1982A and Adie 1985).

Thus, in the century culminating in the mid-1980s, the ethos for providing sport and recreation facilities had really gone full circle. From its use by employers as a means of social control, it has gone through periods of nationalism and welfare to arrive back as a method of social control, although this time directed more by the state than by individual employers. This trend has led Coalter *et al.* to conclude that:

> The issues and concerns underlying political rhetoric have exhibited a high degree of consistency – concern with the disruptive effects of social change, social integration and the strengthening of 'community' and problems of inner city

deprivation have all underpinned an essentially utilitarian view of the content and purpose of leisure provision .... Although the utilitarian aspects were expanded in the 1960s to include a more positive view of leisure as a welfare service, attempting to expand recreational opportunities as a right of citizenship, the status of leisure services is still a matter of some ambiguity (1986, 19).

However, social control in late Twentieth century Britain may be rather more deliberate and sinister than a century earlier. As Hall (1985) points out, 'social palliative' may now be a rather more accurate term, suggesting as it does a negative use of recreation as a last resort when there are no more jobs or training schemes for the unemployed (Jenkins and Sherman 1979). This view is supported by the fact that both the police and the armed forces are also now competing to provide recreation facilities, superficially as a means of communicating with the young and disaffected, but ultimately as a means of controlling them.

However, the more closely the role of sport and recreation in society is examined, the more it is recognised that the undercurrent of control has always been an important element of government policy. Even at the height of the welfare status of provision, between the Wolfenden Report in 1960 and the demise of the Labour administration nearly twenty years later, there were repeated references to the role of sport in ameliorating the effects of boredom and the propensity of the young towards delinquency and criminal behaviour (Wolfenden Committee 1960, Secretary of State for the Environment 1977 and Coalter *et al.* 1986). This has even continued to be the case under Mrs. Thatcher's avowedly *laissez-faire* administration in the 1980s, where the state resources for recreation provision have not been reduced as much as in some areas of mandatory welfare provision such as health, housing and education (Bramham and Henry 1985).

However, whilst facility provision is an important first step in encouraging the nation to participate in sport and recreation activities, studies such the Leisure and the Quality of Life work by the Department of the Environment (Dower *et al.* 1981) have shown that the disadvantaged, or those at whom the provision was targeted, were not likely to be receiving their share of the 'recreation cake'. An example given was the Michael Sobell sports centre in Islington, London, which attracted a largely white, middle class clientele although being situated in a poor, multi-racial community. The same pattern could be witnessed at any number of other similar centres.

In acknowledging the failure of the mere provision of facilities to reach the young, poor, disadvantaged, ethnic minorities or, to a large extent, women, the Sports Council began to move away from supporting facility provision to supporting community leaders who could more effectively bring sport to the people. Whilst there is some parochial evidence that such schemes have been effective in drawing more people into participation in sport and recreation activities, it does tend to beg comparison with the middle class paternalism that originally dominated 'modern' recreation provision over a century ago. Equally, evidence from the STARS scheme in Leicester (Glyptis *et al.* 1986) tends to indicate that even the most intensive community leadership is only capable of retaining the long-term interest of those who are, or become, attracted to sport and recreation activities.

In spite of the relative lack of success of these leadership schemes, or perhaps because no alternatives have been found, the Sports Council, local authorities and the police are continuing to use them. Whilst sports participation is undoubtedly the central core of the schemes, attempts to measure their impact are more likely to concentrate on improvements in social behaviour and community spirit and, in the case of the police, improvements in promoting 'friendship and respect'. This has been accompanied, by the police, with a return to community policing in a further attempt to bridge gaps between the authorities and the community.

A question that arises through all these initiatives is 'why sport and recreation?'; what is so special about sport and recreation that it has repeatedly been used as a method of social engineering? At the most basic level it is probably that it represents a way of gaining access to people's leisure time and then, particularly since sport is so concerned with rules and codes of conduct, influencing the ways in which people think and behave. Indeed, participation in any formal activities tends to reinforce the existing hierarchical nature of society; this is especially so in the espousal of the value of meritocracy and the reinforcement of the belief that discipline, effort and skill are rewarded by success and, moreover, that the existing societal hierarchy is the product of fair competition.

However, it is naively elitist sentiments such as these that have really underpinned the failure of successive governments to adequately plan for sport at any time over the last century. Whilst the aim of full employment was still a realisable goal and whilst there was still a genuine belief that the quality of life was improving for everyone, sport and recreation were a useful medium for reinforcing social values on a populace that had every reason to believe in them. However, when that is patently no longer the case and when the part of society at which these policies have been directed no longer has the prospect of becoming part of the wealth creating classes, why should they be willing to submit to activities organised for them as a palliative for their position in society?

2.6 The growth of commercial provision

The advent of mass tourism was facilitated by the arrival of rail links between the big industrial and population centres and the seaside resorts. This inevitably led to the emergence of the first commercial sector in modern times, involving mainly accommodation and catering, although undoubtedly encompassing elements of entertainment such as circuses and fairs. The importance of accommodation as one of the prime sectors in commercial leisure has remained. Quite apart from the popularity of seaside resorts, the arrival of mass private transport in the mid 1950s allowed people a much wider range of holiday destinations, as well as the ability to make more frequent, shorter trips.

By the 1960s, therefore, tourist accommodation had been extended from the traditional centres to wider regions, such as Wales and the South West of England (Davies 1971 and 1973 and National Farmers' Union 1973). Within these tourist regions, as well as close to large centres of population, a range of leisure attractions had been developed which were becoming increasingly popular. Indeed, it was the advent of the private motor car that led to the opening up of much of the countryside, providing both opportunities for local people (Byrne and Ravenscroft 1989) as well as an unprecedented pressure on the communities and infrastructure (Montagu of Beaulieu 1968).

Since these beginnings, the last 25 years has seen a vast growth in the size and range of the commercial sector. In the country this has involved the development or enlargement of many new leisure and theme parks, the opening of many country parks and stately homes, the conversion of commercial farms into leisure enterprises and, over the last few years, a considerable demand for land suitable for golf course and resort hotel development. The size of this development sector is extremely difficult to establish. In his work on the independent museum sector, Middleton (1990) suggests a rate of growth of about 20 per year (representing perhaps 5 per cent of the total) since 1970, whilst the Royal and Ancient Golf Club at St. Andrews has forecast, perhaps unrealistically, the need for a further 700 golf courses by the year 2000.

The expansion of urban leisure facilities has followed the same rapid pattern. In particular, growth has been experienced in restaurants and catering, health and leisure clubs, facilities for squash and, latterly, indoor tennis, cinemas and themed attractions. Of particular note in urban areas has been the integration of shopping and leisure in new edge-of-town locations (Potiriadis 1990), as well as the development of integrated leisure complexes comprising cinemas, swimming pools, bowling alleys and catering outlets, for example (Sibley 1990).

Whilst there have been attempts to document and categorise the commercial sector in leisure (Roberts 1979, for example), the facilities it provides tend to be, by their nature, unique. As such, they are inclined to complement those provided in the public sector, both in terms of the type of opportunities available and the way in which they are packaged (Chubb and Chubb 1981). Indeed, the ability of the commercial sector to respond quickly to new demands has, to some extent, allowed the public sector to adopt a more passive role. This can be seen in the development of squash. It was initially provided through private clubs for a limited market mostly comprising 25 to 45 year old men (Jenkins 1974). In the period up to the start of the 1980s the commercial sector sought to increase the demand for squash through targeted marketing. This meant that by the time that the public sector had geared up to cater for squash there was a much wider mass market available. With this middle ground now occupied by the cheaper but arguably lower quality public facilities, the commercial sector was able to diversify into the recent health and fitness boom in order to maintain a unique niche in the market.

The commercial sector also has an important role to play in innovation and speculation. Whilst this can lead to expensive failures, such as those witnessed with skateboard parks and, on an individual basis, at Britannia Park in Derbyshire and, potentially, Battersea Power Station in London, it can also introduce new facilities with the ability to attract large numbers of users. This happened with the 'revolution' that has overtaken swimming pools in the past 15 years. Firstly the wave-pool, then flumes, slides and the whole leisure pool concept, have been developed by the commercial sector, initially in the USA and latterly in Europe. As Chubb and Chubb explained:

> It quickly became apparent that wave-pool users stayed longer, returned oftener, and were willing to pay substantially higher admission charges than those paid at conventional pools... (1981, 409).

This inevitably led the public sector into the provision of similar facilities, but with charges much more in line with the commercial sector. Equally, as in the case of squash, it also meant that the commercial sector moved on, by diversifying into sauna and solaria suites, catering and late night opening combined with music and light shows.

In overall terms, therefore, the commercial sector, whether in the USA or Europe, has tended to lead the way in leisure provision, both in terms of facilities and management techniques. Equally, as the public sector has followed the lead, the commercial sector has needed to diversify as well as find niches that are not attractive or suitable for public provision. Ultimately it is this that has fuelled the continued growth of the commercial sector. Thus, whilst some initiatives, such as theme parks, have been so successful that they have become established as part of the fabric of recreation provision, it is the ability of the commercial sector to generate high levels of consumer enjoyment that is paramount in its success:

> Because of its special contributions, commercial recreation plays a tremendously important role in expanding and enriching our recreation environments. Variety is said to be the spice of life; for many people, most of this variety comes from experiences provided by commercial recreation enterprises (Chubb and Chubb 1981, 410).

2.7 The compulsory competitive tendering of local government services

The election of the Conservative government in 1979 effectively signalled the end of the comprehensive Welfare State as envisaged by the policy makers at the end of the Second World War. For the first time in over three decades welfare services came under ideological attack, principally for providing publicly what other countries provided privately (Organisation for Economic Co-operation and Development, Group on Urban Affairs 1986). Since it was never formally part of the Welfare State, leisure provision had, seemingly, rather less to lose than those services that were. However, no public service was immune from the efficiency audits begun in the early 1980s (examples include Department of the Environment 1981, Audit Commission 1983, 1984 and 1987). This meant that leisure provision, sharing as it does many of the same origins and attributes as the other welfare services, also displayed the same major weakness of being unable to justify its provision in anything other than the service it provided. Indeed, it was probably rather more vulnerable to central government scrutiny since sports-related agencies, such as the Sports Council, had very obviously failed to 'establish the political legitimacy of the welfare status of leisure provision' (Coalter *et al.* 1986, 159). Furthermore, its reduction or cessation was hardly likely to be the subject of a major public outcry, especially if there was seen to be a trade-off between leisure provision and health, housing or education.

When compulsory competitive tendering was finally announced as the solution to the perceived inefficiency of government services, therefore, leisure provision was one of those services selected as appropriate (Department of the Environment *et al.* 1987). Whilst local authorities were not required to complete the first stage of the tender of their leisure services until the beginning of 1992, early indications are that considerable interest has been shown in the operation of many public facilities, although principally by council direct labour organisations. Although it is as yet too early to gauge the effect of this change in the style of management, where the private sector has been successful in gaining the tender, the indications are that the gulf between the public and private sectors may be narrower than has been traditionally supposed (Gratton and Taylor 1987). Indeed, compulsory competitive tendering may be providing proof that whilst the original Sport for All campaign that underpinned the expansion of state leisure provision in the post-war period had validity, the way in which it was implemented was not necessarily appropriate; that whilst the policy was concerned with people's needs, the response has been largely concerned with facility provision. Thus the advent of competitive tendering is now removing those facilities from immediate state control, meaning that local government, at least, has been forced to return to the roots of its present policy in order to examine how, in the future, it should ensure adequate leisure provision to meet people's needs.

2.8 Conclusions

As this chapter has shown, the evolution of leisure policy has tended to follow that of society. In the main it has been used either as a diversion or excuse for other changes in society, or as a 'tool' in the process of social engineering. It is, therefore, hardly surprising that no uniform long term policy for leisure has emerged. Equally, it may be mistaken to believe that there are really any fragments of a coherent policy at all. Thus, just as responsibility for the English Tourist Board was shifted from the Department of Industry to that of Employment to suit the new political expediency of the 1980s, the link between physical health and the well-being of society has tended to assume greater prominence in times of social unrest, such as in the industrial revolution of the Nineteenth century or the post-industrial revolution of the latter part of the Twentieth century.

Throughout the century, however, the effect of government action, at both national and local levels, has combined to facilitate the provision of a wide range of recreation opportunities. Since most of this provision has actually occurred in either the private or local government sectors of the economy, central government, for the most part, has been content to underpin that provision through legislation, financial aid and, particularly over the last three decades, a wide range of information and advice. Indeed, this split has tended to reinforce the dichotomy of central government as the generator of investment in the economy and local government, allied to the private sector, as the generator of consumption.

In terms of development and operation this dichotomy is of the utmost importance. For whilst the lead on the provision for leisure opportunity has been taken by central government, the actuality of provision has remained firmly in the hands of those who command control of the necessary resources. Furthermore, since there has never been a policy of land nationalisation, nor of the compulsory acquisition of land for the purpose of providing for leisure opportunity, the ultimate provision of leisure facilities has remained firmly vested in those who own, or can acquire, suitable land. Thus the provision of leisure facilities by local government, whilst part of overall state provision, at least to the extent that those authorities had the power to provide, should really be seen as the response of land and resource owners to opportunities to further the achievement of their own aims. This is really no different a response to that of the private sector, although both the aims and the type of facilities developed may vary.

In essence, therefore, the evolution of recreation provision is really the history of a partnership between central government as policy maker and those who control resources, whether in the public or private sectors, as policy implementors. The measure of the success of this partnership has been the relative strength of the partners and the degree to which their separate aims have coincided. Thus the range of economic and social policies derived by central government have provided rich pickings for both local government and the private sector. However, where the aims have been at variance, over the provision of small local sports centres or the attraction of major international leisure developments, for example, it has been the implementors rather than the policy makers that have determined the shape of provision. In examining the planning and development of recreation facilities, therefore, it is necessary to start with the decision processes of the resource allocators before considering how government can modify or control the nature of recreation development.

3 Property, Property Markets and Development

3.1 Introduction

As is sometimes the case in the English language, specific words and terms, when often used, can take on loose, approximate meanings that do not reflect either full or correct definitions. This appears to be particularly prevalent in the field of property and property development, where references are often made to the ownership of property, rather than interests in property; where the term 'development' is rarely used with precision (Lichfield 1956); and where references are often made to 'the property market' without clarification of what it encompasses. This chapter therefore seeks to clarify some of these issues, with a discussion of property, property rights and property power in Section 3.1, property markets in Section 3.2. and development in Section 3.3. Chapter 4 then describes the process by which land and property are developed, using a sequential progression from development intentions to the actual physical works. Finally, it goes on to consider how developments are, and should be, monitored and evaluated, both during construction and afterwards, in operation.

The term 'property' as in the context of 'the property industry' or 'the property profession' is usually considered to be synonymous with 'land' or at least 'land and buildings'. Indeed, one of the many definitions of property, dating back over 200 years, is that of a piece of land owned, or a landed estate. However, there is an important and fundamental distinction between 'property' and 'land', since the former most properly relates to rights, whilst the latter relates to a physical entity.

The most significant aspects of 'property' are, therefore, belongingness and rights; the condition of being owned by, or belonging to some person or persons, and the right of that person or persons to possess, use or dispose of that possession, whether or not it is land. In contrast, land has a legal definition which includes, according to **Mitchell v Moseley** [1914] 1 ch. 438, CA, 'the surface and all that is [above] – houses, trees and the like ... and all that is [below], ie mines, earth, clay, etc' (quoted in Card *et al.* 1986, 487). Land can also include things that are attached to it, depending upon how securely they are attached, and for what purpose. It also includes vegetation growing naturally on the land, as well as certain incorporeal rights such as easements.

23

In point of definition, therefore, 'land' and 'property' are quite different concepts with the only connection being that 'land' can be a form of 'property'. Over the years, however, this connection has been considered to be highly significant, with 'landed property' forming both a visible as well as calculable form of wealth, power and influence. However, the concept of 'landed property' is not quite as straightforward as it seems, for nobody can 'own' land. Rather, people can own rights and interests in land. That is, people's property consists of rights of possession, use and disposal of defined parcels of land, with the absolute ownership of that land being vested in the Crown.

Land has, therefore, a physical definition and presence which will remain unaltered regardless of political or economic system. It merely refers to a parcel of land, buildings, trees and other physical attributes that can be accurately described by reference to plans, map co-ordinates or other form of physical identification. Land can, then, be considered to be a factor of production, a necessary prerequisite if buildings or other structures are to be developed (see Figure 3.1).

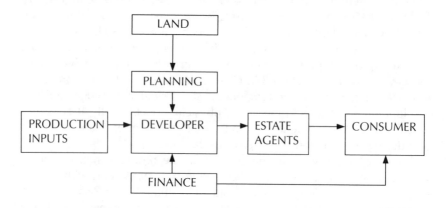

(Source: Drewett 1973, Figure 6.1)

Figure 3.1   Inter-relationship of decision agents in the land development process

Of course, land itself can have properties; that is, attributes or qualities attached to it. These properties can be very important in recreation pro-vision, especially for resource-based facilities such as parks. These properties belong to land and are not, therefore, synonymous with land. Equally, land can have new properties attached to it, in the form of buildings, services and consumers. In the case of an hotel, the complete 'product' (that is, land, buildings and associated properties) may be made up of physical inputs (land and buildings), abstract inputs (staffing, accommodation, catering, consumer experiences) and consumption (see Figure 3.2).

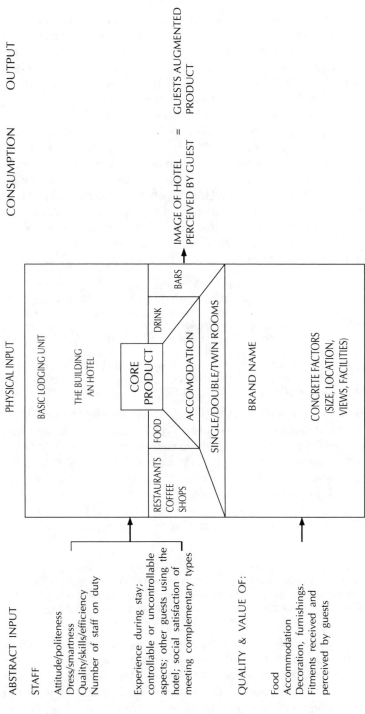

Figure 3.2   The elements of a hotel product

(Source: Gilbert and Arnold 1989, Diagram 1.)

Property itself is, however, essentially a political concept, concerned with the nature and methods of owning and controlling the use of land. In the capitalist systems of Britain, the United States of America and other Western democracies, property rights in land are held by individuals (or by the state in its capacity as an individual). Whilst, in British legal theory at least, outright land ownership is vested in the Crown, we tend to think of the freehold interest in land corresponding to outright ownership.

In collectivist systems land is held by the state, with groups and individuals given the right to make use of the land for certain predetermined functions. Whilst it might be argued that those who gain use rights from the state have acquired property in land, the power of determination is sufficiently limited to make this a meaningless concept. Equally, since landownership is vested in the state the nature of the ownership of property interests is also redundant, since the state does not require the right to an interest before determining the use of the land. As Clark (1982) notes, therefore, private property rights are by nature political rather than natural:

> Generally, property rights are thought to be absolute moral values, independent of the social forces of the community at any given point in time. ... However ... the property rights claim is seriously mistaken on two grounds. First ... what are thought to be absolute rights are in fact relative rights derived from the fabric of society ... Dominance of either individual rights or the community good must be seen as the outcome of existing social relations not some mythical set of natural rights. ... Second, what is thought to be an inherent conflict between individual rights and the community good ... (is) a debatable philosophical issue of how one understands society in general (Clark 1982, 122).

This view is supported by Bonyhady (1987) in his work on the law of the countryside. When dealing with the public's right of access to private land, he shows that whilst medieval people appeared to have more freedom of access than they have today much was *de facto* rather than of legal right. In any dispute over access rights, therefore, the landowner usually won. This was partly because it was felt that *de facto* access would continue, but more importantly because it was simple, convenient and politically expedient to give the landowner exclusive enjoyment of all things on, under and attached to that land. Moreover:

> ... private property and exclusive rights were, according to classical economics, a means of ensuring efficient use of resources. This viewpoint was highly influential from the eighteenth century ..., and may well have formed the basis of much of the common law's preference for private property at the expense of public or, for that matter, local customary rights. (Bonyhady 1987, 9).

Whilst supporting the view that property rights are derived from society, Seabrooke (1989) attributes their creation to the wish to use land as a source of wealth and power. Thus he argues that the Normans created rights to use land distinct from ownership so that power and wealth could be vested in the Crown, with property use-rights being exchanged for military service. This exchange was eventually commuted to cash payments as the mercantile economy evolved. Proprietary rights were also developed, according to Seabrooke (1989), as an acknowledgement of the need to reconcile the physical permanence of land with the individual mobility demanded by society:

> Because 'rights' are capable of being divided, shared and transferred in a way which is impossible for land itself, the emphasis on property rights ostensibly gave land a degree of 'portability' which, otherwise, it so conspicuously lacked. It created a source of wealth and power capable of being assigned or traded but so rooted, so immobile as to be incapable of being stolen away in a covert manner ... (Seabrooke 1989, 3).

The result of this social need to create rights in property that are separate from the land itself has been, in Western capitalist régimes, a division of property rights into ownership interests and interests in another person's land. Whilst the nature and extent of these interests vary between countries, the system evolved in England and Wales is a representative example. Whilst, since the Normans, the absolute title to all land has been vested in the Crown, it is normally considered that ownership interests are limited to two major categories: freeholds and leaseholds. The leasehold interest is synonymous with unencumbered ownership of the land, encompassing the exclusive right of possession, use and disposal subject only to third party rights and restraints enforced on behalf of society. The leasehold interest encompasses a right to occupation and use, often exclusively, which is acquired from the owner of the freehold, or of an existing leasehold interest. The freehold interest is owned in perpetuity, with no rent payable to the Crown or any superior landlord. The leasehold interest, in contrast, will be for a certain number of years, or from year to year. The owner of the leasehold interest will pay a rent to the owner of the freehold interest, or to the owner of any superior leasehold interest in the same parcel of land.

The class of property rights other than ownership interests is wider and includes such interests as easements, covenants and mortgages. Equally, these interests can relate to, or be owned by, individuals or by the entire population. Thus a public footpath across a field which is let by an estate owner to a farmer would involve a right 'owned' by all members of the public, a leasehold owned by the farmer and a freehold interest owned by the estate owner. Furthermore, the contract agreed for the leasehold interest may contain a covenant that the farmer will not build houses on the field. This covenant is effectively 'owned' by the owner of the freehold interest since, subject to planning consent, the owner of the freehold interest possesses the right to build houses on that land. Finally, the freehold owner may have borrowed money to acquire the freehold, via a mortgage. This mortgage gives the money lender a legal interest in the freehold until the loan is discharged.

Under this system of private property rights control of the use of land is largely in the hands of those owning the property rights. The initiation of change, whether by individuals or the state, requires acquisition of the appropriate property rights meaning, ultimately, that the state can only dictate land use by acquiring the freehold of the land concerned, whether by agreement or compulsion. Of course this power of determination is not absolute. The beneficiaries of the non-ownership interests in property described above have rights that must be respected by the owners of those property interests, whilst:

> politicians of all parties accept implicitly some form of the 'public interest' as justifying a ... restriction on private rights in property. (Harrison 1972, 260).

Whilst a market-orientated system of land allocation and control operates in all capitalist economies, there are some notable differences in the allocation of rights that have great significance to recreation provision. One of the commonest of these is the degree of public access to the countryside. In England and Wales this is limited by right to public rights of way and rights of access to urban commons. This is supplemented by a variety of access arrangements to other open land, such as some of that owned by local authorities, the National Trust and the Forestry Commission. Whilst recognising these rights and opportunities, however, Pye-Smith and Hall state:

> We must insist on a right of access to all countryside on foot, subject to no damage being done and the privacy of those who live in the countryside being respected. The 'right of common access' already exists in Sweden, where it works well, and there is no reason to suppose that a similar system would not work here. (1987, 59).

Public access to much open land is maintained in the United States of America by national ownership of all land in the National Parks. This means that whilst some people may have to travel considerable distances to the National Parks, once there they can wander at will, without threat of trespass or encroaching upon someone else's property rights. A similar end is achieved in much of Canada, through state and provincial ownership of large tracts of land. In Ontario, for example, over 90 per cent of the land is publicly owned. Since most of it is coniferous forest, private timber companies can bid for leases, provided they sign Forest Management Agreements which commit them, in addition to felling timber, to replant felled areas, care for wildlife and provide for public access and recreation. In addition, the Province of Ontario retains large areas as public parks, primarily for public access and recreation.

In terms of public policy, therefore, the extent to which land use is dictated by state planning is a function of who holds the property power. Power itself 'is a slippery concept' (Simmie 1981, 7) which generally denotes the ability to produce desired effects. Whilst the most direct method is by ownership of all factors of production, as in a collectivist economy, power can also be exercised through force, coercion, dominance, manipulation or influence in a market-orientated capitalist economy.

Since one of the most fundamental reasons for acquiring property rights is to gain the authority of property power, when seeking to achieve desired effects the extent to which land use follows land policy is politically determined. This is because politics is essentially 'the process by which power, and the different forms of exercising it, is translated into the results required by groups, and individuals with particular and competing interests' (Simmie 1981, 8).

In a capitalist economy, therefore, there are two levels of decision-making over the use of land: government policy and planning; and those who hold rights in property (Denman and Prodano 1972). Whilst the traditional view of these two decision-makers is that the government regulates and the individual initiates change (Harrison 1972), the reality is somewhat more complex, both in terms of who makes what decisions and whether those decisions are wholly innovative or regulatory.

Indeed, in England and Wales, as well as many other countries, the government seeks to protect the rights of those holding property, as well as safeguarding the public interest by imposing controls over the use of property rights in private hands (Denman 1978). Equally, private property rights can be used to ensure that no innovation or change takes place at all, or that the change that does take place is considered retrogressive by the public. Such an example was the case of **Attorney-General v Antrobus** [1905] 2. Ch.188, concerning a proposal to fence off Stonehenge and charge for admission to view it. The proposal caused sufficient outrage to warrant an editorial in *The Times:*

> ... if it is within Sir Edmund's power to enclose Stonehenge with an open fence and to charge a shilling for the right of entry, it is equally within his power to enclose it with a high park paling or a brick wall, to charge a guinea for admission, or to exclude the public altogether. Thus the most complete and impressive specimen of megalithic work in the British Isles ... may be altogether closed to the nation, which had had free access to it from time immemorial. (**The Times**, 20th April 1905, 7; quoted in Bonyhady 1987, 11).

This quotation does, indeed, encapsulate the nature of the power of property rights for, as Denman argues:

> ... to take a decision on how, for one reason or another, the land, water and minerals of a nation shall be used is of no consequence in the ultimate event unless he who takes the decision has power of execution also. The positive power to execute is synonymous with the power to use, dispose of and alienate; the property power. (1978, 38).

The property power is, therefore, synonymous with the ownership of the freehold interest in land. This means that in theory the owner of the freehold interest has the ability to use not only the land itself, but everything above and below that land, subject only to any limitations in the grant of that land or any obligations occurring under common law (Moore 1987). Owners of freehold interests are, therefore, equivalent to absolute owners, with complete freedom to use and dispose of their property rights as they wish.

Over the years, however, the absolute right of use and alienation has been limited, both for practical purposes and to protect the rights of other members of society. In general, these limitations take two forms: limits to the physical extent of the land; and limits to the freehold owner's rights of use. In terms of the physical limits to the land, the right to air surface above the land is limited to 'such a height as is necessary for the ordinary use and enjoyment of the land and the structures on it' (Card *et al.* 1986, 489). Similarly, whilst the freehold extends below the land surface, the Crown retains the ownership of gold, silver, natural gas and oil, whilst coal is owned by British Coal. Thus even the freehold owner requires a licence to extract these resources from the ground and has no right of ownership thereafter. Finally a freeholder can own the water standing in a pond or lake which is part of the land holding. Water flowing on or under the land cannot be owned, but the freeholder does have the right to abstraction. Unless the amount to be taken is small or for purely domestic purposes a licence will be required. Where the licence is not required the freeholder may abstract as much as required from underground channels, but must, by common law, ensure that the flow of a stream or river remains unaltered in quantity and quality following any abstraction (Card *et al.* 1986 give a fuller description).

Control over, and modification to, the use of private land has been apparent since the middle of the Nineteenth century, with the early public health legislation. More generally, however, most societies have recognised the need to restrict certain elements of the property power for the public good and, moreover, to attempt to ensure that the use to which the land is put is determined by the long-term interests of the community rather than as a consequence of the individual property power (Moore 1987). In Britain this has led to the evolution of a range of direct and indirect influences on the part of Government and its agents.

Direct action is largely concerned with legislation. There is a broad range of legislation affecting the rights of those owning interests in land, covering public liability, employment, local taxation and the bylaws created by local authorities. In particular, however, the legislation dealing with land use control has the most direct effect on the rights of those owning interests in land. In general, such controls can cover the prohibition of some land uses, zoning, control over highways, access and services, control over the shape, size and other attributes of proprietary land units and, in some countries, control over the ownership of some types of property and property interest. Whilst recognising the desirability of imposing some public control on land use, Denman (1978) does suggest that since an owner has lost the right to develop land in specific ways, the state has, in fact, expropriated some of the rights or property power acquired with the freehold interest in that land.

In addition to direct controls, successive governments have employed a wide range of indirect actions offering either encouragement or discouragement to develop specific uses of private land. The most common form of encouragement has been financial incentives to private landowners. In the case of tourist development, much grant aid has been available in specified regional development areas. Farmers have recently been offered Farm

Diversification grants to encourage them to develop new enterprises on their farms. Both the Sports Council and Countryside Commission have grant-aided specified types of development, as well as providing valuable technical aid and national recreation and sports statistics. The Forestry Commission is also an important source of grant aid for land owners. Whilst these grants are primarily for forest management purposes, all recipients of grant have to discuss with the local authority the possibility of allowing some public access to their woodland. A different form of financial incentive has been offered to those wishing to develop large residential, commercial, retail or industrial complexes, which has become known as planning gain. In order to gain consent for development, the developers have often agreed to provide extra facilities desired by the community, such as leisure facilities and public services. This has recently been made rather more formal in the Planning and Compensation Act 1991 by the introduction of 'planning obligations', to be offered by the developer as part of a proposed development scheme.

At the same time successive governments have also evolved a series of measures designed to discourage voluntary responses. Most noticeably in the property world is the use of interest rate policies to slow the economy down. This inevitably affects property development and changes of land use. Differential taxation is a further method of discouragement, either by regions or areas of the country, or related to specific land uses, such as second homes.

Thus the system of land tenure in Britain, as in much of the western world, is based on the notion of private property rights protected by the state to achieve a range of national and regional objectives. This means that a balance must exist between those owning property and those seeking to control its use. It also means that conflict can occur when the wider state objectives do not correspond to those of the individuals with property power or to local communities. Equally, it can lead to less than optimal land use allocations when the state can only achieve its objectives by the use of its own resources.

A particular and recent example of these conflicts is the case of government using leisure and tourism as a means of regional development. As recently as 1985 the central government ministry responsible for tourism changed from the Department of Trade and Industry to the Department of Employment. With that change has come a shift in emphasis, so that:

> While English Heritage moves into the tourist business, the English Tourist Board has become a major investor in museums. Capital grants from the English Tourist Board have become a significant factor in the growth of independent museums ... (Hewison 1987, 101–2).

Whilst this investment has resulted in new jobs being created, many of these jobs have been low-paid, transient and temporary and have been largely mopped-up by government schemes such as that underwritten by the Manpower Services Commission (Hewison 1987). At the same time the arts in general has shifted from merit good to externality and, more recently, to being an industry in its own right, created partly for employment and regional development, and partly as a Twentieth century replacement for the power of the church and monarchy, in providing symbols through which the nation and culture can understand itself (Hewison 1987).

Whilst there can be no denial of the wealth generation and employment created by leisure and tourism, the way in which it has been seized upon by various levels of government has resulted in:

> ... top-down planning and promotion that leaves destination communities with little input or control over their own destinies (Murphy 1985, 153).

At first sight this tends to seem contradictory to the notion of individual property power. However, the designation of development areas, the provision of grants and promotion of particular areas can influence the way in which those owning property interests perceive their land and the possible uses to which it can be put. Thus whilst the sanctity of the property power can remain intact, planners' authority over the use of powers lying in the hands of others can bring about change that was neither desired by the community nor directly initiated by the owners of the property power (Denman 1978).

The planners' authority over the power to use land can, therefore, be used in an assertive way on occasion. But this does presuppose that the owner of the property power would take the hint, or seize the initiative, since:

> The prime motive for tourism development and planning has been commercial and economic gain, both on the part of the private sector entrepreneurs and governments (Murphy 1985, 156)

This supposition is not difficult to defend. The same cannot be said for the provision of facilities and services that have essentially non-financial, non-commercial benefits; that are the result of market failure. Many of these services are provided directly by local authorities (albeit managed on occasion by commercial contractors) and include extensive use of land for the provision of leisure and recreation facilities. In these cases of direct provision it might be assumed that positive planning could ensure optimal land allocation to meet community needs. However, as Gratton and Taylor state:

> Sometimes the management of leisure services is to an extent doomed well before the first customer crosses the threshold of a facility. This is because the planning of the service is inadequate. The location of a facility, for example, may be simply in the wrong place for its most likely market, so that its penetration of that market will never achieve full potential. Many local authority leisure service facilities are located not at the best possible site, but at the site that was the cheapest at the time of conception, which in many cases was on land already owned by the local authority. This has a certain short-term logic, but does not represent sensible long-term planning. (1988, 135).

In conclusion, property has been shown to consist of three main factors:

land; rights; and power (see Figure 3.3). Land itself, with any associated buildings and infrastructure, exists in a physical sense that is separate to both people and political ideology. In the extreme it does not require people to determine its use and without the influence of humans will develop a use conditioned by the natural environment.

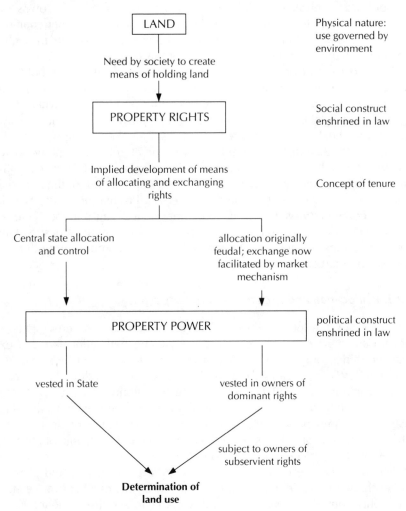

Figure 3.3   Land, rights and power

Property rights are a social construct. They have been developed as a means of rationing and determining the use of land. Property rights themselves require people but not a given political system; they will exist in any system although their nature may be quite different. However by implication the existence of property rights does indicate a means of allocating and exchanging those rights, which depend on the existing political system. Thus, in Britain all property rights are ultimately held by the Crown, whilst in collectivist countries they may be held by the state. In Britain the Crown has delegated its rights, without consideration, to a system of tenure involving freehold and leasehold interests. In colloquial terms the freeholder 'owns' the land and the right to determine its use. Thus in Britain, and other capitalist countries, the allocation and exchange of property rights occurs through a market mechanism developed by people for this express purpose.

The third facet of property is power; the degree to which the holder of property rights can determine or control the use of land. In any economy, or political régime this power ultimately lies with the dominant owner of the rights over that land. In Denman's words, this power 'is the only positive power and the holders of it the only decision-makers equipped to take action' (1978, 32). In the communist state, therefore, the property power rests with the state allowing positive planning at a state level. In the capitalist state, in contrast, the property power is vested in the owner of the freehold, whether this is the state or an individual person. Positive planning, even in cases of community welfare or national importance, can only be achieved by the freeholder so that ultimately the state can only secure a desired land use by acquiring the appropriate freehold, either by compulsion or through the relevant market.

3.2 The allocation and exchange of property rights

Regardless of the political system or the nature of property rights, a mechanism must exist whereby interests can be both allocated and exchanged, since only in this way can the right to use land be conferred. Whilst different political systems and nations have developed their own unique methods, all will lie between the dichotomy of collectivist state provision and capitalist market allocation. At the one extreme the state retains ownership and control of land whilst allocating use-rights to certain citizens or groups of citizens. Thus the allocation and exchange of property rights would occur centrally with no individual having any overriding right to determine the use of land. Conversely, pure capitalism would allocate and exchange rights in a perfect, free market. In this mechanism the state could only control any property that it was able to acquire in the market; it would, in effect, have a status similar to that of the individual.

Whilst the two extremes appear relatively clear-cut, few collectivist economies exist without some element of individual property power, even if limited to some residential or business accommodation, whilst most capitalist markets will be modified by some degree of state manipulation or control. Equally, it might be argued that the same underlying forces are prevalent in any allocational mechanism; namely the relative values ascribed to land in different uses. Thus whilst value in the capitalist market may equate with a market price, and value in a non-market collectivist economy may be more concerned with utility or equity, the forces determining allocation and exchange are both driven by a pre-determined or widely held view of value.

In discussing the allocation and exchange of property rights, therefore, the actual system may be less relevant than the way in which value is ascribed to property and, in the context of this book, where leisure fits in to this process. In the text that follows, therefore, reference is made to the system in Britain with the full knowledge that it is but one example of such an allocational mechanism, but that it may be at least partially representative of what happens in other countries.

The term 'market' whether applied to landed property, stocks and shares, or pigs, signifies little more than that these items are traded; that they can be bought and sold. It says nothing about who the traders are, what rights they actually possess, or even about the location of trading activities. Certainly a market can denote an actual location, such as the Stock Market in London or a local cattle market. These locations are, however, based more on convenience than any prerequisite to trading. The prerequisites of a market are, therefore, buyers and sellers; demanders and suppliers.

The market for property is a highly complex one, based on a multitude of sub-markets. Apart from the most basic division, between property for occupation and for investment, the property market is comprised of three principal sectors: occupation interests that are traded in the letting market; property investment interests; and the development sector.

> Within the three main market sectors, there are an infinite variety of sub-divisions, defined according to property type, location and quality. There is the prime office letting market in the City of London, the prime farmland investment market in East Anglia, the prime industrial letting market in Reading and, doubtless, secondary shop markets in Bootle, Cleethorpes and Clacton-on-Sea. Each sub-market will be subject to its own unique economic conditions, with market operators who would not necessarily consider another area as a sufficiently close substitute.
>
> (Fraser 1984, 121).

In supporting this view, Cadman and Catalano (1983) state that the property market in the United Kingdom has been described as speculative. This is because few occupiers would expect to commission new buildings, but would expect to be able to purchase an occupational interest in a vacant new or existing building. Because of this, the argument goes, property developers must anticipate demand in order to provide buildings that prospective occupiers want. Whilst this is certainly the case in some parts of the property market, such as residential development, and in some areas, such as office and industrial space in the Thames Valley, it is by no means true of the entire property market, since speculative development will only be concerned with providing buildings for which people will pay, either individually, or collectively via the state. It will be further limited to those developments where people are prepared to pay enough to provide the developer with an adequate return for the work (Lichfield 1956). In the cases of market failure, where speculative development will not occur, the collective demand of the state may be necessary to ensure the right buildings are developed:

> The collective demand of a public body for development is usually satisfied by the body itself carrying out the development, so that its problem in forecasting is to gauge whether it, and any other public body which will participate in the development, will elect to spend the money on the development at some time in the future. (Lichfield 1956, 62)

Within this category of market failures are many aspects of leisure provision, particularly those associated with community and social welfare. However, there are other cases in which the concept of a speculative market may also be false. This concerns the importance of site and location to certain types of property, and to leisure property in particular (Mendoza n.d.). The demand for land and property can be divided between a direct demand for consumption services and a derived consumer demand for producers' goods. In the former case consumption services would include houses, schools, hospitals, sports centres and cinemas. In these cases the consumer is concerned with the actual development, both in terms of utility and location. Development of land for one of these uses could be speculative, but the developer would have to be critically aware of the market for it to be a success. In contrast, producer facilities such as factories, offices, shops and warehouses are not of interest to the ultimate consumer. Rather, the interest lies in what can be produced from these facilities. Accordingly, the speculative developer may have a wider or less specific market, so taking some of the risk away from the development.

Due to these types of characteristic, property markets have a number of features that distinguish them from many other markets. These are manifested by competition that is far from perfect, and may approach monopoly for any given transaction, a high relative cost of transfer and some special problems over the management of property assets (Enever 1989).

Amongst the most distinguishing features of the property market is the heterogeneity of property. Every property occupies a unique location and most properties feature different designs, aspects, sites and age. Whilst some types of housing may be interchangeable, the areas over which this is so are likely to be small, due to roads, schools, shops and other services (Enever 1989). Allied to heterogeneity is the fixed location of land relative to society. This can lead to land values changing between locations or over time, which may precipitate changes in the use of the land, even if its location is predetermined (Fraser 1984). Equally important is the stock elasticity. Since the overall supply of land is virtually fixed and changes of use can take time, due to statutory safeguards and controls, the response of the development industry to changes in demand can be slow, so leading to price fluctuations.

Another set of factors that distinguish property are those concerned with ownership and occupation. Since land is durable and buildings relatively durable, the right to use property can be separated from its ownership (see

Section 3.1). However, the indivisibility of land, and the high relative cost of each indivisible unit of land, mean that property purchase will usually result in the use of credit, meaning that changes in financial markets can have a direct influence on property (Enever 1989). This is compounded by the high cost of transfer, where solicitors are required to investigate titles, arrange contracts and secure mortgages, surveyors are required to ensure the quality of property being transferred and valuers are required to advise on pricing and marketing the property.

Finally, the nature of property means that it requires constant management; that it cannot be neglected or simply left in a safe place. Equally, there are a variety of special risks inherent in owning property interests, including physical ones such as fire, flood, earthquake, damage or wear and tear; liabilities to third parties due to defective buildings; the financial risks of credit availability or granting leases; an economic risk of obsolescence; and a political risk of government intervention or control (Enever 1989).

Because of the evident complexity of property and property assets, it is common to think of the market being divided into sectors. Whilst there is no standard list of property market sectors, the main ones are:

i)   shops
ii) offices or commercial property
iii) industrial
iv) residential.

Within these sectors there are, of course, many sub-divisions related to type of property, location, size and so on. In addition to these main four, there is a specialist investment market in:

v)   ground leases

As the name suggests, a ground lease is an annual payment for the land upon which a property is developed. Where ground leases occur they are usually for long periods, such as 99 years. This means that an investor will receive a low ground rent for 99 years followed by possession of the site. Ground leases are occasionally used in leisure-related businesses, particularly for hotels and catering facilities. In using this device a landowner can retain ultimate control of the land whilst allowing someone else to invest money and expertise in developing and operating a business from the site.

Another market sector, sometimes sub-divided, is:

vi) agricultural property and woodland.

Agricultural property itself is usually divided between investments and owner occupation, with the principal investors being wealthy private individuals, the Church and Crown, Oxford and Cambridge Colleges, financial institutions and pension funds. Woodlands have become an investment proposition in their own right due to a period up to the end of the 1980s when tax concessions meant that they were particularly attractive to high tax-payers.

The final sector of the property market is, inevitably, something of a catch-all, known by Enever (1989) as:

vii) special properties.

Within this sector are the properties that are not covered by the other sectors, but are not numerous or important enough to warrant their own sector. These include many leisure facilities such as cinemas, hotels and pubs as well as other properties such as petrol filling stations.

The leisure sector of the property market is further complicated because the overall supply of leisure facilities is unusual in being comprised of three types of supplier: commercial; public; and voluntary organisations. In terms of consumer expenditure, the commercial sector is dominant. Indeed, 'the growth in leisure spending over the post-war period has made the commercial sector in leisure one of the most buoyant sectors in the private sector of the economy' (Gratton and Taylor 1988, 93). The leisure market is, according to Gratton and Taylor (1988), characterised by rapid growth. There is, however, a high risk caused by volatility and a limit to growth caused by a ceiling in demand. Thus many private sector leisure firms have diversified. Typical of this is the Ladbroke Group which has moved from betting shops to hotels, homecare retailing, DIY and latterly, property development and management. Similarly, Bass has moved from brewing to hotels, betting and bingo. Most diversification tends to remain within leisure, partly due to the firm's existing expertise and because they recognise that total consumer spending on leisure is not volatile and has exhibited slow, steady growth over time (Gratton and Taylor 1988).

However, analysis of the commercial sector in leisure is further complicated by a difficulty of classification. This is because not all firms involved in leisure provision classify it as their major venture. Thus whilst the *Financial Times Actuaries Leisure Index* includes 34 listed companies (at 31st December 1989, see Table 3.1) it does not include brewers such as Bass, the Rank Organisation, which is almost wholly leisure apart from Rank Xerox, and other companies such as Pearson, owner of Madame Tussauds, and RMC Group plc, owners of Leisure Sport Ltd, the operator of Thorpe Park (Hilliar 1990).

Traditionally the public sector of leisure provision has been seen as quite distinct from the commercial sector. This gap has narrowed significantly over the last two decades as new indoor facilities have been developed, and since local authorities have become more involved in trading activities through their sports centres, theatres, arts centres, golf courses and similar facilities. Whilst this has meant competition with the commercial sector, the recent introduction of the compulsory competitive tendering of local authority leisure services has led to a new form of public/private cooperation, with firms such as Mecca Leisure managing facilities owned by local authorities.

The effect of the voluntary sector on the property market is hard to estimate since so little information is available. However, club membership

Table 3.1 Financial Times Actuaries Leisure Index
(as at 31 December 1989)

Company	Price (p)	Market Value (£m)	%
Ladbroke Group	341.0	3 056.1	19.0
Forte	311.0	2 438.2	15.6
Thorn EMI	777.0	2 243.4	14.4
Carlton Communications	801.0	1 481.4	9.5
Granada Group	338.0	1 005.7	6.4
Queens Moat Houses	115.0	798.1	5.1
Mount Charlotte	74.0	640.9	4.1
Mecca Leisure	164.0	511.0	3.3
First Leisure	220.0	301.7	1.9
Thames TV	528.0	258.0	1.7
Other 24 companies		2 865.0	18.4
Total:		15 599.5	100

Source: Hilliar (1990)

has been growing fast over the last two decades, with many member clubs owning their own facilities or leasing them from local authorities. In general, however, these clubs do not enter the market for new facilities, except where they wish to expand or relocate to new sites.

Whilst there is little doubt that the property world is becoming aware of the existence of leisure properties, there is still much resistance to investing in leisure, or even taking it seriously (Ratcliffe n.d.). This has certainly been due to the attractiveness of other property investments in recent times, meaning that investors have not needed to stray outside the world they know (Keith 1989). In addition, however, the 'faddish' nature of much leisure activity combined with bespoke, inflexible buildings has further put off potential investors (Chase 1990).

This unwillingness to invest has meant that few comparative performance measures are yet available for leisure property, leading inevitably to difficulties in valuing leisure properties. In considering this problem Mendoza (n.d.) notes that few property professionals appear to be able to differentiate between the value of land for hotels, sports complexes, golf courses and multiplex cinemas, whilst:

> Some of the prices paid for leisure development opportunities cannot be supported in terms of cash flow forecasts. Too many novice investors and developers in leisure appear to be assuming that the rental growth often present for offices, shops and similar forms of real estate, also establishes parallel criteria for leisure property values (Mendoza n.d., 2).

Because of these factors, the leisure industry is often thought to be highly fragmented with too many development schemes being inadequately financed yet over-borrowed (Ratcliffe n.d.) and, consequently, lacking covenant strength (Chase 1990).

These problems are, as Enever (1989) suggests, exacerbated by the location of such properties often having a quasi-monopolistic element. Thus there may only be one cinema in a town, or one major leisure attraction. Equally, in the case of garden centres or roadside cafés, position may be of such importance that even though two competing facilities are in close proximity, the superior position of one means that the earning potential of the two properties is totally different. Furthermore, from a property point-of-view there is an added difficulty when a major element in the value of a property investment relates to specific details over which the investor has no control (Enever 1989). Apart from the location of the property, these details may include the business acumen or other attributes of the tenant or a current fashion in leisure activity (Chase 1990).

Because of these factors the leisure and recreation sector of the property market is currently very narrow, compared to the others, with relatively few buyers and sellers active at any one time (Ratcliffe n.d.). Furthermore, given that leisure and recreation development is capital intensive and dependent upon managerial expertise, high levels of return are required in order to compensate investors. Yet:

> ... in certain sectors of the leisure industry such as golf and squash it is possible to acquire ready-made investments to show a satisfactory yield at less than the cost of constructing a new facility leaving nothing for the residual value of the undeveloped site ... (Ratcliffe n.d.)

It is, therefore, hardly surprising that the market sector is narrow, with the majority of new development being for owner-operation rather than investment.

However, against this background of investor prejudice and misunderstanding there is evidence that leisure property is gaining respectability:

> Although investment yields vary across the leisure market, when compared to yields sought in other sectors of commercial real estate, they are generally high, implying that small capital growth is invisaged. The best leisure developments, however, when backed by management contracts or leased to good covenants, have shown steady and substantial growth ... (Ratcliffe n.d.).

This has now been backed up with the first equity fund devoted solely to leisure starting in 1990 (Terry 1990). With an initial capital fund of £40m, Electra Leisure received over 70 proposals in its first weeks of operation, all based on leisure development.

Finally, leisure is now becoming an integral part of many larger schemes, usually involving retail development. Initially many of these new mixed leisure, catering and retail developments included the leisure element as a means of finding favour with local planning authorities in order to gain permission to develop (Sibley 1990). This can still be the case with developers offering some leisure enterprises that may not be profitable, such as water-based recreation, in return for the right to develop.

3.3 The development process

In general terms development is usually taken to mean the act or process of growing or developing. In terms of the land, development seemingly has a clear meaning, relating as it does to any land to which fixtures have been attached by construction (Lichfield and Darin-Drabkin 1980). With this notion of development comes the verb 'to develop' which conveys the process of 'development' which brings about the end-state of the process, 'the development' (Lichfield and Darin-Drabkin 1980). With respect to the environment and society, Bartelmus defines development as:

> ... a process that attempts to improve the living conditions of people. Most also agree that the improvement of living conditions relates to non-material wants as well as physical requirements. Development goals that call for the increase of human welfare or the improvement of the quality of life reflect this agreement (1986, 3).

Current legislation on the control of property development takes a more limited view of what constitutes development. Under s. 22 Town and Country Planning Act 1971 development can take two forms: 'the carrying out of building, engineering, mining or other operations in, on, over or under land'; or 'the making of any material change in the use of any buildings or other land' (Moore 1987, 56 quoting the 1971 Act).

Beyond its legal definition, however, McNamara (1983) sees the development of land as a direct result of development decisions coupled with an investment of labour and capital in land. For development to exist, therefore, the following criteria must be satisfied:

i)   there must be a material change in the use of the land;
ii) there must be an 'investment' of land, labour and capital;
iii) there must be a purpose behind the act.

(McNamara 1983, 88)

Purpose is also seen to be a key element by Cadman and Austin-Crowe (1983), with the primary motivation behind any development being the provision of accommodation for occupation by the developer or a third party. However, as Lichfield comments:

> The word 'development' is used very widely in relation to land but, outside certain Acts of Parliament, it is rarely used with precision. Among architects, engineers and surveyors it generally means the process of carrying out the constructional works which are associated with a change in the use of land or of land with its buildings, or with a change in the intensity of the use of the land, or with a re-establishment of an existing use (1956, 1).

Beyond these broad definitions of development, however, there is an apparent lack of detailed information. The reason for this, according to Cadman and Austin-Crowe (1983), centres on the reluctance of traditional developers to record or recount their attitudes and methods. However, it is apparent that there are some systematic or objective elements within property development, particularly relating to when development occurs, who the developer is and who the development is for. Since the property market in Britain is depicted as being speculative, developments will commence as opportunities occur:

> The development process begins when a parcel of land is considered suitable for a different or more intensive use ... (Goodchild and Munton 1985, 65).

This suitability can become apparent from a number of sources. Barrett *et al.* (1978) suggest that a range of influences can 'trigger off' a move towards land conversion. These influences range from changes in public sector planning and land use policies to private responses to changing external circumstances. Drewett (1973), however, attributes the market with being the prime determinant of development, since the purchase of land constitutes both an investment decision and a locational one. However, Drewett goes on to point out that underlying all these influences is the one central factor of the decision made by the developer:

> The urban development process is the aggregate outcome of many decisions in a complex social-economic-political system, which will remain difficult to articulate and comprehend, unless a deeper understanding is achieved of the decisions determining it (1973, 163).

In terms of the process of development, most authors tend to see it as a sequence of events following a logical, staged course from beginning to end. The beginning and end, as well as the stages themselves, do tend to vary according to the interests of the authors. Thus, in the work on aesthetics by Punter (1986), two of the three stages of development are associated with initiation and obtaining planning consent, whilst four of the six stages in the work on industrial development by Fothergill *et al.* (1987) are concerned with events before construction itself occurs.

Perhaps the most general view of the stages of the development process are outlined in Cadman and Austin-Crowe (1983) and again in Cadman and Catalano (1983). These are:

i)   Evaluation
ii) Preparation
iii) Implementation
iv) Disposal (or use).

Evaluation would involve the identification of a demand or need for new buildings (Evely 1986). It would also involve seeking suitable locations and actual sites for the proposed development. Evaluation of the development cost, likely return and site value would provide a broad indication of feasibility.

Provided the evaluation stage produced positive results from the feasibility study, preparation for development would commence. This would involve detailed designs, arrangements for funding, applications for planning permission and other statutory consents, acquisition of the land and assembly of any sub-contractors or other labour force. The length and complexity of this stage is highly variable, depending upon the time taken to agree a design, gain planning consent and purchase the land. Equally, variations or restrictions in the planning consent can render the original plans subject to total redesign (Punter 1986 gives a full description of this phase).

Once all the preparatory stages are complete, construction can commence. Included in construction would be the assembly of all materials on site, the arrangement and staging of the finance to cover the construction period and a provision for clearing and landscaping the finished site. If it was intended to dispose of the development upon completion, the last stage, involving marketing the building, would also begin at this time, if not before. Disposal could involve the freehold of the site, but could involve retaining the freehold and letting to tenants. Finally if the developers have pre-sold or let, or will occupy the building themselves, the final stage of use will commence at the end of the construction phase.

In essence, therefore, this view of the development process suggests two models with a common core. The models relate to whether the development is speculative or for a known client. They will differ only in the first and last stages, where the criteria for initiation will vary as will the need for marketing and disposal. However, these stages, and particularly the first of them, are vitally important to the process, since the remainder are largely technical in nature. This has led, as McNamara (1983) points out, to a wide range of attempts to model the development process. However, since few have sought to focus on the purpose of a development and the relationship between developer and potential occupant, few models have been able to do more than describe a sequential, technical process.

It is, perhaps, this concentration on describing the technical process of development that has contributed to this apparent lack of emphasis on the purpose of the development and the nature of the developer and occupant. Nowhere is this more apparent than with leisure facilities. The process of gaining statutory consents, preparing a site and constructing a building will not vary greatly between private and public sectors, yet the type of building and its locational requirements may vary considerably. These differences will be due largely to the purpose of the development. The public sector will be responding to a need by local residents for leisure facilities (Mercer 1973), whilst the private developer will be responding to a perceived gap in a particular market. These differences will influence the type of facility built, as well as its location.

In recognising the drawbacks of the technical approach to modelling the development process, McNamara (1983) developed a classification of developers according to the purpose of their developments. This has been modified in Table 3.2 to relate the purpose of development to leisure property. Although somewhat over-simplified and subject to discussion, if not disagreement, Table 3.2 does illustrate that the developer and purpose of development are more important than the type of development or its construction.

Table 3.2 Classification of developers by purpose of development

Before development	Ownership of interest in land		
	After development		
	Short term	Long term (lease out)	Long term (own and occupy)
Short term	Entrepreneurial builder – gap in market exploited by large development company. Examples include timeshare, hotels, pubs and restaurants; – may be part of planning gain for a new non-leisure development	Land developer/investor – large property company, pension fund or insurance company wishing to extend portfolio. – high street catering, pubs and hotels are common	Developer/user – farmer or landowner seeking to reinvest under roll-over provisions. – leisure operator having recently acquired site – eg Rank leisure parks, multiplex cinemas and outdoor 'theme' parks.
Long term (lease out)	Asset clearing, probably investment switch – large investment company/fund recognises opportunity to dispose of freehold at the end of a lease. Often for high income ventures such as restaurants, clubs or hotels	Property improver/rentier – property company renovating leisure facilities before re-letting on new terms and rent	Expanding developer/user – public sector obtaining possession of site and re-developing for its own use – landowner taking facility in hand to redevelop and operate
Long term (own and occupy)	Capitalising assets – private landowner converting traditional land/buildings to a new use before sale	Change in returns from property – private owner of golf/squash/hotel hotel/pub etc decides to let to new tenant for an annual rent instead of seeking an operational profit – local authority subject to compulsory competitive tendering	Owner-occupier/designer – local authority building a new facility to meet public need – farmer seeking to diversify into non-farming venture, often accommodation, pick-your-own or recreation

Source: from McNamara (1983)

This, in itself, brings into question whether the development process is most appropriately modelled in technical terms at all. This is recognised by Cadman and Catalano (1983) who, whilst using technical models for description, suggest that the development process might more appropriately be seen as a financial or a socio-political process. The argument for a financial model rests on the basis that the development process is concerned with the creation of assets largely for investors. This is particularly relevant for property concerned with the prime office and retail markets. However, even within leisure property markets there is still considerable activity by investors and property owners seeking financial returns that compare to other forms of property and investment. Indeed, it might be argued that only in the long-term occupation sectors of Table 3.2 does finance become less than a dominating factor, and then only for public sector operators.

Closely linked to the financial view of the development process is the socio-political view, where developers and landowners are seen as using development to perpetuate inequality in society, rather than simply as a means of generating income and capital:

> According to this (*socio-political*) view property development is not a 'neutral' process but part of a system in which political and economic power is unevenly distributed, and where those who control the means of production benefit at the expense of the rest. Thus, development is geared to perpetuating the existing inequalities in society. (Cadman and Catalano 1983, 29)

This view of development, whether termed socio-political or merely political, is pertinent to all development, but particularly so for leisure property, along with other types of property associated with public welfare and provision by both public and private sectors.

As detailed in Chapter 2, central government has been intervening in the leisure facility market throughout the last 100 years in order to modify the allocation of resources to certain types of provision. Implicit in this was a utilitarian view that leisure provision was a means of improving society rather than being an end in itself and, allied to this, that the national government saw itself as a legislator and facilitator, leaving the actual provision to local government and the voluntary and commercial sectors of the economy.

Recent affirmation of 'the three aims of Government policy in sport' (Moynihan 1987) as improvement of the nation's health, alleviation of social deprivation and the promotion of excellence, whilst at the same time questioning the continued provision of leisure facilities by the state at all (Department of the Environment *et al.* 1987) acts as a reminder of the fallibility of such provision, when it is predominantly carried out under permissive, rather than mandatory powers, by local authorities. Thus, although leisure opportunities have been viewed as part of a vague set of social rights, the extent to which the public sector should subsidise or provide, to compensate for market failures, is uncertain. This is due, on the one hand, to an enduring social ideology of leisure being a private individual concern and, on the other hand to the continuing wish by government to use leisure provision as a means of social manipulation and control; this inevitably means that leisure services in Britain 'occupy an uneasy place between ideologies of the market and ideologies of welfare' (Coalter *et al.* 1986, 159).

Regardless of the degree to which successive governments have provided for social services, the basis of the British economy has been the competitive market. Under this system the individual entrepreneur has been encouraged and provided for, particularly by the continual enhancement of the 'private capitalist' value system as the favoured basis for decision-making. Because of this, much is known and has been written about the capitalist market and value system, and, in the context of supply decisions, many processes and techniques have been developed to aid management and investment decisions (Lumby 1984).

Equally, provision for basic welfare services, such as health care, continue to be accepted, largely regardless of the political ideology governing the management of the national economy. The provision for such services may increasingly be by means of subsidy to the private commercial sector, and any welfare service that remains directly provided by the state may be reduced to the lowest standard of service deemed acceptable by government (as is currently the case with the National Health Service). However, as long as axiological values based on the worth of health, education and minimum living standards continue to transcend political ideology, the state provision of such services is ultimately assured, regardless of any ability to measure or justify their worth in the economy.

What is much less clear or certain is the fate of sectors of the economy that feature goods that are neither wholly private nor wholly public, as in the case of leisure provision. If left to the market there would be under-provision of facilities and opportunities, whilst the public provision of leisure facilities is coming under increasing ideological attack from central government in Britain, due to its peripheral welfare status, and the fact that it is provided through the market in some other parts of the world (Le Grand 1982 and Organisation for Economic Co-operation and Development, Group on Urban Affairs 1986). Whilst the justification of successive government's leisure provision has been based on the enhancement of social policy, therefore, the ideological attack has largely been in the form of detailed financial scrutiny. This has, in effect, introduced evaluation criteria unrelated to the objectives of provision and based on a value system different to that upon which the welfare services were originally provided. Maynard (1983) stresses this, with respect to the National Health Service, by stating that if its main aim were financial efficiency it would not be structured in the way it is at present. Examples of this type of service can be found on the periphery of Welfare State provision, where public provision was probably established under more welfare-orientated ideologies, but these services have less claim to be fundamental social rights than the triumvirate of health, housing and education. This type of provision can be termed 'quasi-welfare' and includes much state provision for recreation, sport and the arts.

By inference, the types of service that fall within the quasi-welfare category have never been completely considered as 'social' services in the sense of being indispensable services to provide for those that the private capitalist market will not. Rather, they tend to be services with strong elements of welfare attached to them, but retaining significant elements of private capitalism, and remaining largely discretionary in their social context. This quasi-welfare sector cuts across the capitalist/collectivist dichotomy by containing substantial elements of both value systems. This means that whilst being largely 'social' by nature, it does not tend to be either politically or socially indispensable. Furthermore, unlike purely private or purely public goods, the motives for provision and, hence, the output expected, are unlikely to be capable of evaluation under a single value system, but are likely to contain elements of private capitalism and public welfare. To some extent, therefore, such services might be expected to produce financial returns to the provider, as well as a non-financial output, or return, related to the motives for provision, but incapable of evaluation in similar terms to the financial returns.

In addition to the political nature of public sector leisure provision, the private sector is following no less a political path. When use or consumption is discretionary, as in the case of leisure, the private sector can make a decision about the type of client they might wish to attract, knowing that exclusion of some members of society may be a positive benefit to their intended clients. This manifests itself in many ways, with high entry fees and restricted membership of private sports clubs, new out-of-town centres for leisure and retail catering predominantly for car owners, and the 'up-grading' of many pubs, restaurants and hotels, with attendant price increases.

In the countryside this trend is just as prominent, particularly where farmland and buildings are put to new leisure uses. Where previously people might have walked or ridden horses, even without formal legal right, golf clubs and new owners of converted barns now seek to exclude them, not only from the land but also from any part of the activity. Equally, landowners can reclaim the use of redundant barns from let farms, modify them and relet or sell them with little benefit to the existing farm tenant.

Even in cases where private developers may claim to be providing public amenities within commercial developments, these 'parasitic' facilities (English Tourist Board and Jones Lang Wootton 1989) are often the result of planning gain – the 'cost' of obtaining a planning consent. However, whilst the local authority might be able to determine the type of facility built, it will not be able to determine the location, nor ensure that future management will be appropriate to local peoples' needs.

In conclusion, therefore, the property development process is by its nature political in context; and the politics concern choice and function of the development. The process must, therefore, start from the values and ideology of the development instigator, whether or not they undertake the actual work. Once the development has been initiated the process becomes technical in nature, following a path through evaluation, preparation and construction. It then resumes its political nature in terms of its disposal and subsequent use. These elements will now be developed more fully in the following chapters.

4 The Development Process and the Consequent Demand for Land

4.1 Demand, need and the development process

Since the basis of any development decision must be founded in the concepts of consumer demand or assessments of societal need, one of the fundamental considerations in analysing the development process is to consider the relationship between demand, need and recreation development. In simple terms demand is concerned with the willingness and ability to pay for a desired good or service. In this context, need is represented by a collective demand by the state, financed largely through tax revenues, for goods and services deemed essential for the wellbeing of the community served. The initial problem for developers, therefore, is to assess or forecast demand prior to development. However as Lichfield states, with respect to developers:

> Their problem differs according to whether the demand is individual or collective, direct or derived; and to whether they are building in speculation for unknown customers, or by contract for known customers (1956, 60).

Consumers are said to have a direct demand for facilities when the goods produced yield consumption services; that is to say, when the consumer requires access to the building or facility in order to consume the goods produced. In addition to common buildings for which there is a direct demand, like houses and schools, many recreation developments fall into this category. Such developments would include sports halls, swimming pools and cinemas. There is a derived consumer demand, conversely, for buildings or facilities yielding producers' goods, where the consumer is not interested in the building, but in the output of that building. The most common examples of these buildings are factories, offices and warehouses. Whilst few of these types of building could be classified as recreation developments, many are necessary to support recreation activities, such as sports equipment factories (see Chapter 3 and Lichfield 1956).

Whilst, in the eyes of the developer therefore, the consumer demand for recreation facilities is direct, to the consumer it is derived from the direct demand for recreation activities, but conditioned by some facility-specific variables such as the attitude of management (Gratton and Taylor 1986). This is particularly important, in that many popular recreation activities, such as fishing, walking, jogging and cycling do not require the provision of any specialist recreation facilities, whilst in those that do, the developer may have little or no control over the attitude of management. Equally, demand by the state is derived, to the extent at least that it is based on perceptions of need rather than willingness or ability to pay.

Forecasting the demand for recreation facilities is, therefore, a complicated undertaking, but one which underpins the remainder of the development process and, eventually, helps determine the developer's demand for land upon which to construct the development. As outlined in the last chapter, the development process itself describes the stages whereby land is transformed from one use to another. This chapter will seek to expand on that description by considering the principal stages in detail. Section 4.2 will, therefore, consider the intention to develop, including the aims of the client or developer and the identification of appropriate markets; Section 4.3 will expand the discussion of how different providers and developers undertake feasibility studies; Section 4.4 will deal with the financial appraisal of development projects; Section 4.5 will consider physical development itself; whilst Section 4.6 will outline the methods of evaluation suitable to different developers and operators.

4.2 The development intention

In the early literature on the development process, such as Lichfield (1956), Drewett (1973), Barrett *et al.* (1978) and Lichfield and Darin-Drabkin (1980), little attention was given to the development intention, with rather more stress laid on individual parcels of land and the circumstances under which development would occur. Whilst not refuting this orientation, Goodchild and Munton (1985) suggest that rather more weight should be placed on those making the development decisions; the participants or actors in the process. And that of these actors, the developer is considered to be the most important in initiating the development process.

Goodchild and Munton (1985) suggest that there are three principal types of development project: those for sale; those for letting; and those for occupation by the developer. As they go on to state:

> Development of all three types is carried out by both the private and public sector, but the amount of public sector development for sale is limited. Different factors are important for public and private sector developers. Public sector developments are not normally undertaken solely for financial reasons. A local authority may make a profit from carrying out a central area re-development scheme but the scheme is undertaken with the main aim of improving facilities for the local community. A private sector developer is primarily seeking a profit from a scheme, although prestige obtained from carrying out a particular development may be a subsidiary consideration. Public and private sector developers tend to approach their tasks differently. The public sector developer proceeds more slowly because of the need to consult the public and to have decisions approved by political masters. The private sector developer attempts to complete the task as quickly as

possible, unhampered by such consultation unless imposed by public bodies ... (Goodchild and Munton 1985, 68).

However, whilst a simple description such as this might suffice for speculative developments and improvements to existing developments, recreation and leisure facility development is less straightforward. This is partly due to the mix of facility and management that sets leisure buildings apart from many other types of development, partly due to the existence of the voluntary sector as a third type of developer, and partly because the majority of leisure development is destined for occupation by the developer.

In the case of leisure and recreation development, therefore, intention becomes a concept of central importance since it will distinguish the different reasons or motives for provision and, therefore, the parameters governing the success or suitability of the development. A number of authors have distinguished motives for the provision of leisure facilities, including Davison (1988), Field (1988), Gratton and Taylor (1988), Hewison (1987), Hilary (1984), Murphy (1985), Roberts (1990) and Sibley (1990). Whilst most of these authors are more concerned with particular motives than an exhaustive list, the following seven major motives emerge, each of which will be considered below:

i)   statutory commitments and responsibilities;
ii)   income generation;
iii) capital growth or gains;
iv) regional development and employment generation;
v)   planning gain and grant assistance;
vi) club or membership wishes; and
vii) altruism, public image and public service.

4.2.1 Statutory commitments and responsibilities

Whilst a great many government departments, at both central and local level, have some responsibility for leisure provision (Blackie *et al.* 1979), most state-initiated recreation development has been carried out by local authorities, partly in response to their duties under the Education Act 1944 and partly in response to their community responsibilities enshrined in local government legislation. Examples of the former include school sports halls and playing fields, often open to the public as at the Hurst School near Basingstoke, whilst examples of the latter include facilities such as the leisure centres, swimming pools, parks, gardens and golf courses found in most towns and cities (see Table 4.1 for an indication of the nature and extent of current local authority provision). On occasions, central and local government have combined to provide facilities which satisfy both local and wider (national and/or international) needs, such as the National Water Sports Centre at Holme Pierrepont, Nottingham.

Table 4.1 The range of publicly provided facilities

Category	Facility
Sport and recreation – outdoor	Playing fields Golf courses Bowling rinks Stadia Marinas Ski slopes
Sport and recreation – indoor	Swimming pools Gymnasia Sports halls Ice rinks Leisure centres
Informal recreation – mainly outdoor	Play spaces Amenity open spaces Urban parks Beaches, lakes, rivers
Countryside recreation	Country parks National parks Camping sites Picnic sites
Cultural recreation	Concert halls Theatres Art centres Art galleries and museums
Education-related recreation	Adult education centres Youth clubs Community centres
Library services	Branch libraries District libraries Mobile libraries
Tourism, conservation and heritage	Information services Historic sites Nature reserves Conservation areas
Entertaining, catering and conferences	Public halls Pavilions Conference centres
Housing, community and social services	Play centres City farms Allotments Day centres Community halls Holiday camps

Source: Torkildsen (1986)

Generally, public leisure provision has been undertaken on the dual basis of allocative efficiency and social equity. The efficiency criteria are based on the existence of externalities and the public good aspect of some leisure provision; equity criteria are more closely based on ideology, but relate to a belief by successive governments that some aspects of leisure, notably active recreation, are 'good' for the members of society. For the last 25 years, the basis of this state intervention in the provision of leisure and recreation has been loosely organised around the Sport For All policy first identified by the Wolfenden Committee on Sport (1960).

Although never codified in a policy document, and having been added to over the intervening years, the underlying ideology inherent in the Sport For All concept has become the basis upon which the state has developed facilities to provide for leisure, recreation and sport. The full range of the Sport For All initiative has been collated and analysed by McIntosh and Charlton (1985). In outline, the Sport For All policy is based on six major aims:

1) to increase participation and improve performance in sport;
2) to have sport treated as a social service;
3) to produce a range of social benefits, such as the maintenance of moral standards and improved social welfare;
4) to produce a range of psychological benefits, such as the enjoyment of leisure and the advancement of personality;
5) to produce certain physiological benefits, such as an improvement in the nation's health and fitness; and
6) to improve the quality of life of the nation.

In association with the six policy aims are nine specific objectives, designed to work towards the fulfilment of those aims. The objectives are:

1) to promote participation amongst certain target populations;
2) to provide for gifted performers;
3) to provide and optimise the use of facilities;
4) to help ameliorate certain social problems, such as boredom and delinquency;
5) to encourage commercial investment in sport;
6) to promote sport for all;
7) to reduce coronary heart disease;
8) to encourage research; and
9) to collaborate with other nations in sport.

In outlining the Sport For All policy in such a way, it must be recognised that the aims and objectives are not a code established by the state and handed down to all public facility developers. Rather, they include some statements of policy, some initiatives and some actions from which the aims and objectives can be deduced (McIntosh and Charlton 1985). Furthermore, it must be recognised that the Sport For All policy relates to sport and active recreation, but does not necessarily cover all the aspects of leisure for which facilities are developed. In particular, the Countryside Commission, in expressing state policy for the use of the countryside, takes a more low-key approach to provision (Countryside Commission 1982).

Thus, the broad aims of the state in developing facilities to provide for leisure, recreation and sport involve a wide-ranging set of initiatives based on the implicit assumption that increased participation will lead to an overall increase in social welfare. The result of these initiatives was that capital expenditure by local authorities began to rise in the early 1970s as new indoor facilities were developed. This continued well into the 1980s, in spite of the new central administration. Equally, recurrent expenditure also increased since most of the new facilities made a financial loss (Gratton and Taylor 1988). Whilst public leisure development may remain small and selective, when compared to the private sector, leisure developments are now a major element in local authority budgets, as well as being at the centre of the recent debates on the compulsory competitive tendering of leisure services management.

Other areas of recreation development covered by statutory commitments and responsibilities include support for tourists, in the form of tourist information centres operated by local authorities and interpretive centres run by park authorities within the national parks. English Heritage is also involved in recreation development in providing car parks, visitor centres and access to historic monuments, whilst local authorities may provide car parks as part of an agreement for the public to gain access over private or open land.

4.2.2 Income generation

There are two major types of facility developed to generate income: those where income generation is the primary aim of the operator, as in the case of the large publicly-quoted leisure firms; and those where income generation is secondary to, or supports, a separate primary activity, whether commercial or not. Examples of the latter include: shops and catering facilities developed at local authority sports and cultural facilities (often including a bar and restaurant, as at the Harlow Sportcentre); extensive shops, cafes and restaurants at the large national museums, such as the Tate Gallery, London; and a wide range of catering and retail outlets at historic houses, whether in the public sector (such as Hampton Court), the private sector (such as Chatsworth), or the quasi-public sector (such as Kingston Lacey House, owned by the National Trust). Equally relevant to the latter aim is the current trend of developing leisure facilities as an incentive or complement to retail outlets, such as at the Metro Centre, Gateshead.

The former type, where income generation is the primary motivation, is synonymous with the commercial sector of the leisure market. This has a great variety of forms, covering recreation, tourism, sport and the arts. It also covers those actually providing participatory opportunities, equipment manufacturers and distributors, the tour and travel trade and those in the hotel and accommodation trade (see Figure 4.1).

The commercial sector of leisure is one of the fastest growing and most buoyant sectors of the UK economy, as well as being dominant in the supply

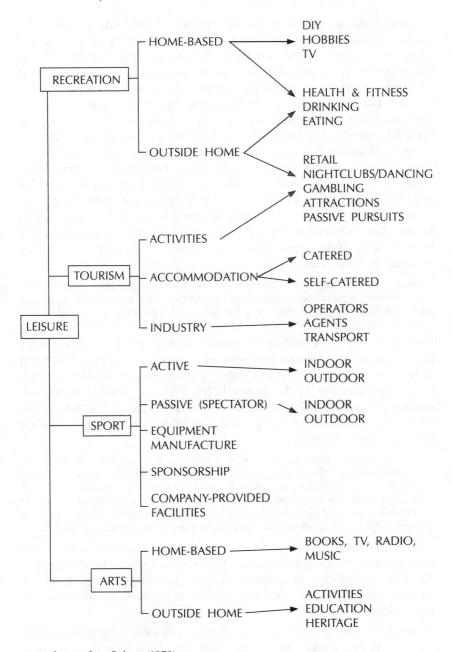

Source: from Roberts (1979)

Figure 4.1 Commercial sector involvement in lesiure

of leisure opportunities. It is also the sector that most conventional economic theory relates to (Gratton and Taylor 1985). That is, the provision of goods and services for the purpose of deriving a profit. Thus the primary determinant of performance is likely to be return on capital invested, with the goods and services yielding the greatest net revenues being preferred.

Whilst Gratton and Taylor (1988) suggest that in the case of commercial leisure provision the economic model of profit maximisation might be replaced with a satisfactory performance criterion, they do point out that commercial operators are not interested in sport and recreation *per se*, not because it is a merit good or because society is likely to be a better place if people are active in sport and recreation. Rather, they supply recreation goods and services because they are likely to become richer as a result (Gratton and Taylor 1985). This does not mean to say that the commercial sector in leisure has no worth beyond that of any commercial operation, as the wide range of development intentions covered in this section indicates. However, as Roberts states:

> Arising from the commercial sector's profit-making activities are major contributions to employment and, in some cases, to the export drive. Beyond this, it could be unrealistic to ascribe altruism to the commercial sector to any greater extent than it may be found normally in individuals and groups (1979, 31).

Because of its income and profit orientation, but perhaps contrary to the image of commercial leisure provision, the commercial sector is essentially conservative in its outlook. It will tend to provide what it considers, or what it can persuade, the public to want, regardless of whether technically superior alternatives may exist (Sibley 1990 provides an example of this within large multinational companies). The upshot of this conservatism is well observed by Roberts:

> The commercial sector provides leisure services that, on the one hand, are frequently more efficiently managed at less cost than those of the public sector, and large sections of the leisure market are uniquely met by the commercial sector. On the other hand, this sector eschews activities that are not profitable and is wary of innovation. A significant proportion of the population either cannot reach, or cannot afford, or is apprehensive about using many of its facilities (1979, 39).

One of the prime results of this has been, in the commercial sector, the growing incidence of leisure facilities provided as a result of planning gain. This has enabled the community to gain new largely unprofitable or commercially unattractive facilities without recourse to tax revenues (see 4.2.5 below).

However, the primary motivation of income generation does not preclude other non-financial goals. In his early work on private recreation provision on American farms, Bevins (1971) notes that the decision to provide facilities or services, whilst justified in financial terms, was frequently the result of a complex range of aims and objectives. Similarly, the Dartington Amenity Research Trust (1974) identified four non-financial motives for farmers introducing tourist or leisure enterprises on their farms. These were:

a social motive, related to companionship; interest, where the activity might take the form of a hobby; altruism, in the form of a wish to share the countryside with others; and using the new enterprise as a means of retaining more members of the family on the farm (Byrne and Ravenscroft 1989 provide a fuller discussion of these factors).

Indeed, it would appear in many cases that, whilst the stated intention behind a new leisure enterprise may be income generation, the operator will not even know if the enterprise is actually profitable. In her survey of farm diversification schemes, Paice (1988) notes that only 38 per cent of the sample knew if their schemes were profitable, whilst only 10 per cent enthused about the extra income generated by the new enterprise. More generally, Hillary (1984) notes that one of the main factors holding back a wider range of commercial leisure development:

> ... is the difficulty in achieving an acceptable rate of return on investment to compensate for the very considerable risks involved and management skills required (Hillary 1984, 102).

Once again the effect of this is to restrict commercial firms and entrepreneurs to a narrow range of activities that are known to be profitable. In the case of the Rank Organisation this generally means a mix of leisure, retail and catering, with leisure only included if it is profitable or if it has the ability to generate business for other profit centres (Sibley 1990). Thus, cinemas, snooker halls and bowling are currently profitable and would be included in a development scheme on their own merit. Other activities, principally associated with water, sport and recreation, are not generally considered profitable and would only be included in a scheme because of planning gain, generous grant aid or the ability to generate a lot of business for other parts of the development (Sibley 1990).

The notion of leisure facilities being used to draw customers for other outlets is becoming well documented, particularly where it relates to shopping (Potiriadis 1990). The essence of combining leisure and shopping is complementarity. This implies creating a destination or centre where people will be attracted by the combination of shops, catering and leisure activities available. According to Potiriadis (1990) this can take two forms: a shopping centre enhanced by up to 10 per cent of the area devoted to leisure, as is the case at the Basingstoke shopping centre; or a dual shopping and leisure centre, with the leisure element comprising at least 25 per cent of the total area, as at Tower Park, Poole.

However, experience gained from some of the established developments, such as the Metro Centre, Gateshead, indicate that it is as yet rare for the leisure uses to be able to match the retail uses in rental terms. The leisure element is justified, therefore, in terms of providing an identity for the centre, generating employment (see 4.2.4 overleaf), creating a desired ambience, and widening the retail catchment area.

In other sectors of the economy leisure activities are used as a direct income generator, providing funds to achieve other aims. An example of this is the National Trust, where income from recreational activities account for an increasing proportion of the Trust's annual income. The combination of membership and admission fees has risen from 35 per cent of annual income in 1970 to 51 per cent in 1990, with property rents falling by a corresponding amount (Baring 1991). If product sales and the net contribution from National Trust Enterprises (the Trust's trading arm) are included in total recreational income, the proportion of the Trust's income derived from recreation rises to 58 per cent for 1989 and 60 per cent for 1990 (Baring 1990 and 1991). The largest proportion of this income is used to preserve and maintain the Trust's stock of historic buildings and landscape.

Similar steps have been taken in the private sector of the 'heritage industry', with an increasing number of historic house owners opening their homes to the public in order to generate income for property maintenance. Figures produced by Miles (1986) show that the majority of house opening ventures show significant levels of net income before maintenance outgoings, but relatively few are able to cover all outgoings from recreational income. However, since the houses are usually family homes rather than pure investments, any recreational income must help offset the running costs of the house.

Other examples of income generation to aid the achievement of non-financial primary aims include admission charges to some art galleries and museums, and the 'commercialisation' of retail and catering outlets in local authority leisure facilities in order to generate the greatest possible income to help offset running costs. Indeed, the operation of many local authority services is currently being reorientated towards income generation, given certain political controls, by the introduction of the compulsory competitive tendering of many services. By this mechanism the government hopes that income generation will assume more importance in the public sector than hitherto.

4.2.3 Capital gains and growth

In many ways an intention to develop leisure facilities on the basis of capital gain or growth is similar to the motives for income generation. There are, however, two areas in which the two may not be synonymous, although income generation could be one of the results of capital growth.

The first of these areas concerns the large Public Limited Companies where capital growth may outrank income generation as a corporate goal. Indeed, the *Financial Times Actuaries Leisure Index* (see Table 3.1) indicates that this has already occurred, with the largest five companies accounting for more than half the sector's capitalisation. Since the commercial leisure sector is characterised by rapid growth to a pre-determined ceiling of demand (Gratton and Taylor 1988), the growth goals of the large companies are to be expected, together with constant diversification and take-over activity to ensure their continuity of growth.

At the other end of the scale, many land and estate owners use leisure development as a means of generating capital gains for investment, if not growth. Typically this is achieved through the sale of an existing asset and reinvestment in new assets with different attributes (Byrne and Ravenscroft 1989). There is often some conversion of the assets prior to sale, such as obtaining planning permission for a change of use and imposing covenants on the future use and appearance of the asset.

Given the long-term nature of land and estate ownership, with strong familial and social ties to the land itself, the sale of assets is neither a common nor a preferred strategy. Where it does happen, therefore, the actual area of land involved is usually small and the capital value highly significant. Examples of this would include the conversion and sale of barns and cottages or the sale of small plots of land for development (Byrne and Ravenscroft 1989).

Whilst there is evidence of the sale of such assets by estate and land owners, there is rather less evidence of what is being purchased to replace them. In some cases the capital can do no more than pay off outstanding debts; in others it may be reinvested on the estate, or in new assets outside the estate. In one particular case, quoted by Byrne and Ravenscroft (1989), an estate sold land to purchase an hotel. Apart from the new source of income, it was also felt that the hotel could be successfully tied in with commercial sporting and leisure activities on the estate itself.

4.2.4 Regional development and employment generation

In much the same way that we are inclined to see development as a neutral technical process, so recreation and tourism are seen as no more than vehicles of individual self-expression. But what may be self-expression to one individual can also be part of a larger process of conditioning and change; as Hewison states:

> A display in a museum may simply be telling a story, but the existence of a museum has a story to tell (1987, 9).

This assertion is equally applicable to the case where recreation and tourism can be one person's business or livelihood at the same time as being a government policy. This was certainly the case with the origins of farm tourism in Wales. The original stimuli for introducing new enterprises to farms appear to have been the potential opportunities for tourist provision created by the booming economy of the 1960s, together with the recognition that the economic pressures facing farmers were likely to increase rather than diminish in the future (Davies 1971). The Wales Tourist Board, in lending its support to increased farm tourism, saw it principally as a tool for regional development where benefits would extend beyond increased individual farm incomes by helping to check rural depopulation, maintain adequate or improved services and preserve the traditions, customs and culture of Welsh rural communities (National Farmers' Union 1973).

Not all authors concur with this dichotomous explanation of this facet of recreation and tourism. Indeed, Murphy (1985) states that, as far as providers are concerned, the prime motivation to provide for tourism development has been commercial and economic gain, on the part of individual providers as well as government. However, the nature of government dictates that, in this context, tourism and recreation have been used as agents of economic regeneration in depressed areas, as well as an export activity (Murphy 1985). Equally, within those regions:

> Cities frequently look on tourism as a way to bolster their retail and service sectors, and derive greater returns on their cultural and recreational investments (Murphy 1985, 156).

Examples of this can be seen in many cities, both in Britain and elsewhere. Indeed, even in an era of solid anti-collectivism, certainly in national political terms, one of the first developments in regenerating London's docklands was the London Arena, built with public monies to attract attention and interest in the docklands. Davison (1988) notes the success of similar approaches in Glasgow, Liverpool, Birmingham and Bradford, where considerable public investment in the arts has led to new confidence and life in these cities:

> The arts, the argument goes, can offer a significant number of jobs in areas where the disappearance of traditional industries has led to high levels of unemployment. They can create more jobs through attracting tourism, and, as other people find a place attractive, they can create a new mood of "confidence". Most important, they can transform the image of a depressed area as a means of attracting new industries (Davison 1988, 28–9).

Within the euphoria of economic development, however, questions must be asked of the effect on the inhabitants of these 'developing' areas, as well as those tempted to participate in, or consume, the new recreational opportunities. Hewison (1987), in particular, is concerned with the imposed effects of such developments. His concerns are, on the one hand, with the exploitation of local people by the use of culture and education to train them for revised economic and social expectations; and on the other hand, by the way society is becoming conditioned to the effects of culture and tourism:

> I call it the 'heritage industry' not only because it absorbs considerable public and private resources, but also because it is expected more and more to replace the real industry upon which this country's economy depends. Instead of manufacturing goods, we are manufacturing heritage, a commodity which nobody seems able to define, but which everybody is eager to sell, in particular those cultural institutions that can no longer rely on government funds as they did in the past.(Hewison 1987, 9).

Britain is not alone in exploiting recreation, the arts and culture. When The Walt Disney Company Ltd announced its intention to develop a new European park most governments recognised its economic potential. The park was eventually sited close to Paris after the French government agreed to build all connecting roads, extend the metro system and cut the tax on ticket sales, in the knowledge that over 10 000 jobs would be created as a direct result (Newman and Roberts 1989). In addition, however, the government went further:

> Over the objections of local farmers, French public authorities used their right of 'eminent domain' to sell Disney and partners a piece of prime real estate in the fast-growing region of Marne-la-Vallée. What is more, they sold the land to Disney at a fraction of market value (Newman and Roberts 1989, 37).

4.2.5 In exchange for public monies or development rights.

One of the most contentious issues in the development of major urban areas in recent years has been the euphemistically termed concept of 'planning gain'. At its crudest, this has involved commercial developers 'buying' the right to develop property on a greenfield site by agreeing to provide facilities for the public – which are often associated with leisure and recreation. Sibley (1990) suggests, from the experience of the Rank Organisation, that local authorities often want water or ice facilities, neither of which is normally commercially viable. This was certainly the case in Reading, where the Rivermead Leisure Pool was built for the community as part of a package including an hotel and an industrial park. In cases such as these, the developer will have to include some element of cross-fertilization in the proposed development package.

In his work on the financing of leisure development, Roberts (1990) sees planning gain as a spin-off from general development gains. This suggests an element of reason for the public to be grateful, or at least aware that the developer has put something of public benefit into the development. Indeed, there is evidence of some local authorities, such as that in Bracknell, which take planning gain most seriously and, as a consequence, have been able to build up a range of publicly available, but privately funded, leisure facilities. In this way such returns from development can, or perhaps should, be seen as a form of betterment levy (Field 1988). If this is the case, Field (1988) argues, local authorities should seek to extend it significantly, particularly in getting the private sector to provide and maintain public spaces, especially in closed settings such as shopping malls. A recent example of this is the Meadowhall shopping centre outside Sheffield, completed in 1990 at a cost of £230m, where the developer was required to provide the Don Valley Linear Park, connecting the shopping development to the city, at a cost of £5m, as planning gain.

Whilst it must be recognised that a great many facilities have been made available to the public in this way, and at nominal direct cost, the efficacy of this type of action is highly questionable. Whilst the direct costs may be minimal or negligible, little is known of the hidden or indirect costs, particularly in terms of lost opportunities to use the resources in other ways. How can local authorities be sure that the public gets the facilities it needs and in the most appropriate locations? Equally, how far can a local authority realistically dictate what is to be built, and where?

Beyond these practical issues of matching public needs to the spin-off of private development, there remains the philosophical question of whether it is defensible to confuse the pressures of commercial development with issues of social provision which are of no immediate concern to the developer, given that the principal beneficiary of planning gain is the developer. If, in pragmatic terms, the future of public leisure provision is dependent upon exchanging facilities for development rights, why not institute a standard betterment levy on all development, which could then be used by local authorities to provide the right facilities in the right places to meet local community needs? However, it remains import-ant to recognise that regardless of whether the system is defensible or equitable, planning gain has resulted in the development of many non-commercial leisure facilities for public use but at no direct capital cost to the public.

Beyond the provision of leisure facilities in order to gain development rights, recreation and leisure opportunities can be made available as a result of government grants and allowances. Whilst not involving substantial new development, some of the principal reasons for opening historic houses to the public include the need to fulfil the qualifications for conditional exemption from capital taxes and because opening is a requirement follow-ing the receipt of grant aid towards the cost of house repairs (Miles 1986). However, even in these cases there will be a need to make provision for public access to the house and, in most cases, the wish to provide associated facilities such as shop, catering and toilets. The number of houses and grounds to which this applies is not known, but since Miles (1986) indicates that the majority of enterprises in his survey were unprofitable, it might be fair to imply that reasons unconnected with profitability determined that house-opening would go ahead.

A similar, although much less onerous condition used to be applied for private woodland owners in receipt of Forestry Commission grants. Under the old Forestry Dedication schemes (Forestry Commission 1979), owners agreed to:

> ... accept a continuing obligation by Deed or Agreement of Covenant to manage all their woodlands within the scheme in accordance with Plans of Operations designed to secure sound forestry practice, effective integration with agriculture, environmental safeguards, and such opportunities for public recreation as may be appropriate (Forestry Commission 1979, 3).

In practice this agreement tended to amount to very little as far as public recreation was concerned, with few owners doing more than the statutory minimum of maintaining public rights of way. However, even this minor commitment to recreation was reduced in 1981 with the introduction of the new Forestry Grant Scheme (Forestry Commission 1981), where a far greater emphasis was put on timber production and the expansion of private forestry in Britain.

4.2.6 Voluntary sector club or membership wishes

The voluntary sector of the economy is both large and diverse, including as it does all sorts of groups from working mens' clubs to youth groups and sports clubs (Tomlinson 1979). The history of such voluntary groupings is long, certainly extending from Greek civilization. Indeed, as Tomlinson (1979) points out, clubs are characteristic of social structure rather than of time or place, and are based upon:

> ... a common general process incorporating a) the fragmentation of the individual's total existence and the atomistic nature of modern social life, b) the individual's articulation of a chosen interest and c) the urge of the individual to give corporate expression to this interest (Tomlinson 1979, 2).

The term 'voluntary sector' or 'voluntary organisation' covers a wide range of groups and clubs, from those with a national presence, such as the National Trust and the Royal Society for the Protection of Birds to village sports and hobby clubs. This wide range of organisations led Hookway (1984) to assert that the voluntary sector has made 'a remarkable contribution' to countryside recreation in recent years. This is particularly so given the broad cross-section of interests covering preservation, conservation provision and strong political lobby.

Whilst membership numbers for voluntary clubs are hard to establish in any consistent manner, it would appear that the vast majority are comprised, at a local level at least, of very few members. Because of this their needs tend to be small, with few doing more than meeting regularly in rooms or premises rented from others, often the local authority. However, some types of group do have a demand or need for a larger, bespoke or permanent base. This is so of many sports groups which require pitches, courts or courses of a given quality on a regular basis.

Whilst many such clubs are still able to rent the facilities they require from local authorities, even in the case of such extensive land users as golf clubs (Johnson 1990), many more have invested, or are in the process of investing, in facilities. Evidence of the types of club which do invest is scattered at best. However, work by Ravenscroft and Stabler (1986) indicates a link between 'social' sports, such as tennis, rugby and golf and facility development. This is born out by the Sports Council, which established that more than half the amateur rugby clubs in Yorkshire owned their own facilities (quoted in Tomlinson 1979). In contrast, few amateur football clubs in Greater Nottingham owned their own facilities (Seeley 1973).

Beyond these broad statements, however, there remains a lack of quantitative information about the activity or operation of voluntary clubs. Because of this it is not possible to do more than note that many do develop facilities and those that have may have amassed substantial capital assets.

4.2.7 Altruism, image and public service

Finally, in this categorisation of development intentions, is the class of facilities developed for altruistic, service or image criteria. In common with the intentions of voluntary clubs, there can be no single or straightforward description of the forms that this can take. However, the role of altruism cannot be overstated, since many of the first 'public' leisure facilities, in Victorian times, were provided by wealthy altruistic members of society. Whilst the care of most of those facilities has now passed to local authorities, there remain many examples of leisure facilities provided for largely altruistic reasons.

Important amongst these facilities are the country parks and pleasure grounds opened to the public at little or no cost by the owners of large rural estates. The 'facility' as such may constitute little more than a right of access, some areas for parking and, sometimes, the provision of toilets. However, such areas can be popular destinations for family outings and do allow the public access to private country over which they have no legal rights (Hilary 1984). Interesting and diverse examples of this can be found at Elvaston Castle, Derbyshire; Rufford Abbey, Sherwood Forest and Holme Pierrepont, Nottinghamshire, whilst the provision of a country park at Goodwood, West Sussex also serves as a means of controlling access to other parts of a private estate, by enabling the estate staff to direct visitors away from private areas to the country park.

Some leisure facilities are also developed in order to enhance the image of an individual, group or company. This can be particularly so where developers wish to prove that they can be trusted, or where firms wish to show that they will give things to the community without the compulsion of a planning agreement. An example of this is given by Hartwright, with respect to gravel pits at Chertsey, owned by Leisure Sport Ltd, a subsidiary of RMC Group plc:

> ... Ready Mixed Concrete was anxious to find ways of developing at least some of these sites on a more commercial basis and at the same time create areas for public enjoyment which would be helpful to the company in demonstrating that worked out pits can be turned to advantage and provide much needed public amenity areas (1984, 58).

In overall terms, therefore, there is a wide range of motives underlying development intentions, whether in the public, private or voluntary sectors. What is equally apparent is that such intentions may encompass more than one particular motive and that there can be no automatic association of motives with particular sectors.

4.3 Feasibility

4.3.1 Introduction

Having discussed the various motives for providing leisure facilities, this section seeks to examine the process by which providers decide whether, and what sort of facility, to develop. Feasibility is essentially, therefore, the link between a provider's motivation and the potential customers or users of the facility to be developed.

As the previous section indicated, the motives for developing leisure facilities are wide ranging. In all cases, however, the providers, although not necessarily the developers, are drawn from three sectors of provision covering the public sector, private commercial sector and private non-commercial, or voluntary sector. Whilst each of these sectors exhibits unique characteristics, all are drawn together in leisure provision, as shown in Figure 4.2.

From Figure 4.2 it can be seen that each sector has its own aims or goals, with the private sector having commercial provision, the voluntary sector having its own initiatives and the public sector deciding to provide some facilities directly. In addition, the public sector forms a variety of associations or links with the other sectors, by providing grants, loans and other incentives such as land and statutory consents. The result of this is a group of five motivations that guide the provision of facilities in all the sectors of the economy:

i)   welfare benefits;
ii) investment for economic growth and employment generation;
iii) investment for commercial gain;
iv) pecuniary benefits; and
v)   non-pecuniary benefits.

The first two of these groups relate to the public sector, with the welfare benefits category covering the multitude of facilities provided by local authorities for the public in general and certain sectors of society, such as schools, in particular. The second category, of economic growth and employment generation, may be initiated by central or local government and may involve association with the private commercial sector. The private sector is also involved in investment, but normally for individual commercial gain rather than regional development or employment generation. Finally, the private sector and voluntary sector will be seeking to achieve pecuniary and non-pecuniary benefits respectively.

In turning to the criteria used to judge the likely success of a particular development, the concepts of demand and need are relevant. Demand is a commonly used term which, when applied to recreation, is often taken to mean the total number of visits to a facility. However, it is more correctly a measure of volume, used to relate factors such as the number of visits to a facility with the price of a visit (Clawson and Knetsch 1971). Demand is also related to people's aspirations, in that demand can exist for things that are currently unattainable, either through lack of appropriate supply or the existence of social, economic or psychological barriers to consumption (Miles and Seabrooke 1977 and Pigram 1983).

Demand is, therefore, an expression of individual wants and the degree to which those wants can be satisfied. Need, conversely, is an institutional

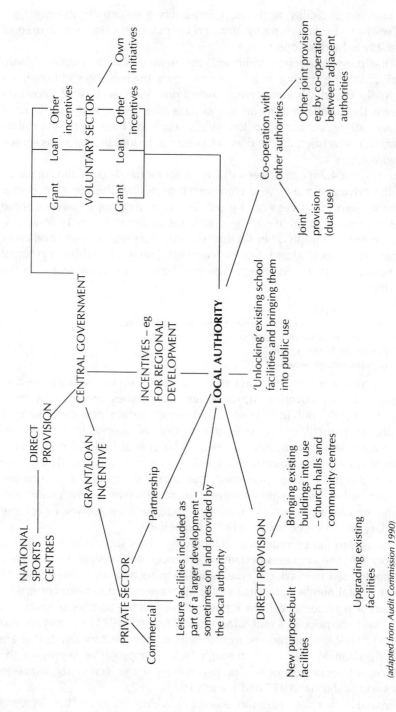

Figure 4.2   Leisure facility provision

(adapted from Audit Commission 1990)

recognition of relative deprivation between individuals and groups in society, regardless of their expressed or implicit wants. Thus demand is essentially a private motivation that can be met through the operation of a market mechanism, whilst need is a public motivation which can only be addressed through the provision of appropriate facilities to rectify perceived or relative injustices. In essence, therefore, the private commercial and voluntary sectors are motivated by the wish to cater for people's demands whilst the public sector is motivated by the wish to fulfil people's needs. The link between the sectors happens largely when the public sector estimates that demand-motivated provision by the private sectors can indirectly meet some of people's needs.

In considering feasibility criteria, therefore, the central concepts to consider are those of demand and need, and the ways in which they influence the process of determining the feasibility of individual recreation developments.

4.3.2 Need

Faced with the seeming ability to provide whatever level or range of leisure facilities is deemed appropriate, a number of approaches have been developed to aid the measurement of feasibility and subsequent provision of such facilities by local authorities. Coalter *et al.* (1986) suggest that most authorities adopt one of two approaches: rationality or morality. The former of these is essentially reactive, responding to social inequalities as they become apparent. The latter is more normative in approach, deciding upon what criteria equity should be assessed and then setting about achieving the desired level and type of provision. Alternatively, LeGrand (1982) suggests that equity provides the basis for all welfare provision. These measures include: equality of public expenditure; equality of final income to all members of society; equality of use according to need; equality of cost to all members of society; and equality of outcome for all individuals.

However, in adapting these approaches to leisure provision, a number of authors, such as Veal (1982) and Field and MacGregor (1987), have found that the methodology available is limited, both conceptually and through lack of data. What does emerge from their work is that need, rather than equity, is the major determinant in the most common approaches to planning for leisure in local authorities in Britain. Mercer (1973) distinguishes four types of need, named: normative; expressed; felt; and comparative, with each having provided the basis for methods of resource allocation in leisure provision.

Normative need is based on the formulation of precise objective standards by experts in fields associated with leisure, recreation and sport (Mercer 1973). The use of such standards is common in local authorities and is based on their earlier tradition of paternalism, allied to the continued inadequacies of forecasting techniques (Field and MacGregor 1987). Examples of commonly used standards are given in Table 4.2.

Table 4.2 Examples of UK standards in planning for leisure

Facility/Service	Standard	Body responsible
Playing fields	6 acres (2.4 ha) per 1000 population	National Playing Fields Association (NPFA)
Allotments	0.5 acres (0.2 ha) per 1000 population	Thorpe Committee
District indoor sports centres	1 per population of 40 000 to 90 000 plus 1 for each additional 50 000 population. (17 m^2 per 1000 population)	Sports Council
Local indoor sports centres	23 m^2 per 1000 population approximately	Sports Council
Indoor swimming pools	5 m^2 per 1000 population approximately	Sports Council
Golf courses	1 9-hole unit per 18 000 population. 700 new 18-hole courses by the year 2000.	Sports Council Golf Development Council
Libraries	1 branch library per 15 000 population. Maximum distance to nearest library in urban areas 1 mile. (1.6 km) Book purchases: 250 p.a. per 1000 population.	Department of Education and Science
Children's play	1.5 acres per 1000 population	NPFA

Source: Veal (1982).

Most of these standards are wholly unresearched and based on the 'hunches' of experts (Bacon 1980). However, wide use and acceptance has given them normative qualities over time. The main advantage in the application of standards is that action is likely to follow, due to their simplicity, efficiency (no duplication of research or data collection), authority (they are deemed to be normative and objective) and measureability (Veal 1982). The endorsement of the standards approach to provision has been more complete than might have been expected for such a system largely due, in the opinion of Field and MacGregor (1987) to the lack of other suitable methods and, in the opinion of Bacon (1980), to its appeal to recreation planners and professionals as the articulators of societal wants.

In response to the criticism of the standards approach as purely empirical, the Organic approach has been developed. This approach represents a method of establishing how to justify and plan the development of facilities using empirical evidence, market research and normative statements about people's needs (Veal 1982). Ultimately, however, any approach based on normative need must rely on the formulation of expert opinion on the type and range of facilities to supply; the greater the evidence collected, the more the approach will be based on objectivity, rather than on intuition, but ultimately all such approaches will be subjective in their application.

Expressed need is based on measures of actual participation in leisure activities and consumption of leisure facilities by individuals (Mercer 1973). It is the basis for the widely used Gross Demand approach to provision (Veal 1982). In this approach the results of national trends and surveys, such as the national countryside recreation survey (Countryside Commission 1985), are applied at the local level. This can either be done in a straightforward manner by applying all the results without qualification, or it can be stratified, by the use of market segmentation, for example. In common with the standards approach, however, it is still empirical by nature. Furthermore, it is questionable how far the results of a national survey can and should be applied at the local level.

A reliance on expressed need can also lead to what Bacon (1980) terms diffusionate planning, where leisure facility providers assess what is popular at present (in terms of consumption or participation) and simply provide more of it. Indeed, it is arguable that this form of planning has underpinned the growth in the provision of sports centres and squash courts in Britain, where current and recent past trends in participation have formed the basis of future forecasts.

The use of felt need as a basis for leisure provision has gained much popularity recently, due to its use of latent demand as the indicator of need (Mercer 1973 and Field and MacGregor 1987). Many local authorities now undertake attitudinal surveys of their resident population in order to ascertain latent demands for different types of leisure activity and facility. Furthermore, a wish to extend public participation in the planning process has led to the use of Community Development approaches to provision, where resources are allocated to the community for the community to decide how they should be used.

Another method is the Grid Approach, which combines felt and expressed need by attempting to combine data on facility usage with attitudes about future demand and consumption. This method is supposed to let planners know which areas or groups in the community are being served by which facilities; and the extent to which the needs of these groups are being met by current provision (Veal 1982). However, such approaches have drawn criticism for their assumed rational basis for resource allocation:

> ... recreational planners ... assume that they act in a rational, detached, scientific manner. They have the professional competence and, because they have done their research, detailed information necessary, to make informed judgements about the provision of the people's leisure facilities (Bacon 1980, 13).

Although this view was published some time ago and the education and training of recreation planners and managers has progressed since that time, it is probable that this is still the domain view adopted in planning leisure provision.

The concept of comparative need is based on the equity of spatial provision or equity of access to facilities, discussed by LeGrand (1982), or relates to measures of relative deprivation (Field and MacGregor 1987). Some comparative approaches have been developed from this conception of need, largely based on bookkeeping approaches and being similar in many respects to normative approaches. In particular, the Spatial Approach considers existing facilities elsewhere, their use, catchment areas and participation rates and applies this to the locality in question. A refinement of this approach is the Hierarchy of Facilities method, which attempts the same process, but simultaneously for a range or group of facilities rather than for isolated individual ones (Veal 1982). Both these approaches were used extensively by the former Greater London Council to ensure equity of provision amongst London Boroughs.

Finally, one attempt has been made to combine all four concepts of need into one method of feasibility evaluation, known as the Issues Approach. It was developed in the mid-1970s, at the time when leisure provision was most closely allied to welfare policy and when the newly formed Regional Councils for Sport and Recreation were campaigning for comprehensive regional recreation strategies. The approach is based on the ability to define issues which are developed from principles, empirical reviews, attitudinal surveys and an assessment of priorities. As such, it is more an approach to conceptualising areas of policy than a method of allocating resources. In this context it has been used, whether implicitly or explicitly, to develop wider planning policies by many local authorities (Veal 1982).

To conclude, the discussion of these approaches to feasibility throws considerable light on the underlying values and processes by which resources are allocated. On the surface, some of the methods appear to be based upon rational processes and, hence, appear capable of measurement and justification. A good example is the enduring use of standards. However, not only is this justification based on the flimsy reasoning that such a standard has been used many times before, but it also fails to connect the authority's motives for provision with the outcomes of that provision. With the possible exception of the Issues Approach, none of the methods seek to explain why certain facilities should be provided, or why certain resources should be allocated and, even more fundamentally, what the community stands to gain by this provision. As Bacon states:

> The domain assumption appears to be that there is a commonly shared, or agreed, series of values concerning what kinds of provision are held dear, or seem worthwhile. Certain concepts such as the goodness of education, a countryside recreation experience, the arts, physical activity, sport and so on, are seen as, or are presumed to be, true ... (1980, 12).

Intuitively, therefore, leisure professionals and planners treat the public interest as a normative goal; and their own jobs and professionalism as those of deciding the means and resources to achieve these pre-ordained series of goals or values. As Torkildsen (1986) notes, the public provision of leisure is traditional, historical and institutional; it is a result of what exists, what local government is geared up to handle, and what is known and understood. Because of this reasoning and the approaches consequently adopted, local government must ultimately fail to know whether it has achieved its goals, even assuming that those goals are appropriate for the community being served:

> ... planning in the field (of recreation) tends to be a highly regressive activity, which probably does as much to allocate more public resources to the 'haves', and to take away from those who have not, as to fulfil the generally democratic, if somewhat piously egalitarian objectives which characterise the official aims and objectives of our ... leisure agencies (Bacon 1980, 17).

In conclusion, whilst political ideology may be the catalyst behind state leisure provision, at the local level, where the majority of public provision occurs, values and ideology are largely replaced by a normative conception of community interest and goals. This is so even where it is manifestly not the case. Conflict can exist between different groups within the community, whilst central government can impose value judgements, or at least monitoring and control, of local welfare provision on the basis of a dominant value system that runs contrary to the normative values assumed by the officers and elected members of the local council.

4.3.3 Demand

Following the broad and somewhat subjective nature of need, the concept of demand appears to benefit from the certainty associated with a widely used economic term. However, as Pigram observes:

> ... there is an apparent inability to distinguish between the concept of demand in the broad, generic sense and its use to refer to the existing level of recreation activity (1983, 16).

In other words, demand has more than one common meaning. The first refers to the number of visits to a facility. This is more properly a measure of consumption rather than demand. The second meaning is the more correctly defined economic concept of quantity related to price. However, since many recreation facilities have no admission price and others a highly subsidised one there can be no uniform cost to users in these cases, meaning that the economic definition of demand may have little application to these recreation facilities. This has led to demand being associated with consumption, with little or no reference to price, even where visitors must incur costs; as Miles and Seabrooke state:

> The rigour of the economist's concept of demand, though valuable for analysis of precise, clearly definable matters, may be less useful in dealing with complex problems which, when dismembered into simple components, lose the inherent complexity around which the problem revolves (1977, 49).

Equally, the aggregate demand approach of the economist may have limited application to an individual site, where the consumption orientated expression of demand is more relevant. This is largely because recreation is a commodity created and consumed at the site or facility. Thus the closeness of the manager-visitor relationship is apparent, so necessitating managers to be aware of individuals' behaviour and the likely influences that exist on their recreation demand. These influences are summarised in Table 4.3.

Table 4.3 Influences on demand

1. Factors relating to potential users as individuals:

    a) the number of people in an area
    b) their geographic location
    c) their average income and the distribution of income
    d) their average leisure time and its distribution
    e) the age and sex distribution of the population
    f) the socio-economic characteristics, such as occupation, education, knowledge of facilities and experience of activities
    g) individual skill, motivation and preference.

2. Factors relating to the recreation facility or area:

    a) the attractiveness of a facility or an area in terms of the quantity and quality of facilities
    b) the management of the facility or area
    c) the capacity of the facility or area
    d) the availability of substitutes or alternatives.

3. Factors relating to the interaction of (1) and (2) above:

    a) the distance or time or cost of travel between the residence or place of work and the area or facility. There is a pronounced 'distance decay' effect which varies substantially for different activities. This is similar to shopping but whereas, in some sense, shoppers must shop, people need not participate in a particular leisure activity. Thus, as distance increases not only does the proportion of the population using a facility decrease, but the frequency of use of individual users also decreases.
    b) the marketing of the area or facility.

Source: Field and MacGregor (1987).

Whilst there may not be much dispute about the list of influences on demand, it remains difficult to identify any structure of causality. Equally, the parameters are often highly unstable over time, with countless examples to be found of changing fashion, such as the shortlived popularity of skateboard parks, the changing fortunes of ten-pin bowling and the increasing interest in sports such as squash and golf.

This situation indicates that at present there is a considerable gulf between our knowledge of existing and potential patterns of recreation demand. The public sector response, based on fulfilling needs, has been to provide basic facilities where people are left to their own devices, thus creating their own enjoyment rather than relying on the site or facility to do so. The private sector, conversely, has tried to bridge this gulf of knowledge by linking its entrepreneurial skill at facility provision with the marketing techniques necessary to convert latent demand to effective demand (Miles and Seabrooke 1977). This latter approach, now being increasingly adopted by the public as well as the private sector, is, assert Miles and Seabrooke 'a serious attempt to probe the facets of ... demand' (1977, 55).

Demand assessment on the basis of market research and segmentation relies heavily upon the ability of management to identify and target specific sections of the population. Indeed, it can be argued that the market for a recreation experience may be of more significance to management than the site or facility itself, since only by attention to the demands of the market can visitor numbers be assured or improved.

In order to develop or maintain this 'people' orientation, the manager must be in the position to determine a marketing strategy. The major decision to be made prior to developing the strategy is which potential visitors to encourage. This process of identification and selection is known as market segmentation:

> At the one extreme of absolute market differentiation it may be said that each individual visitor constitutes a significant market segment by the virtue of the uniqueness of his needs. At the other extreme lies the policy of undifferentiated marketing where the peculiarities of individual visitors are ignored, attention being given to their common qualities in an aggregate market approach. Not surprisingly these extreme policies seldom occur (Miles and Seabrooke 1977, 56).

The degree to which market segmentation can occur depends upon three conditions, as defined by Kotler (1984):

1) measureability: the level of information available on visitor characteristics.
2) accessibility: the degree to which marketing can be concentrated on chosen market segments.
3) substantiality: the degree to which market segments are large enough to be worthy of separate consideration.

Specific difficulties are associated with the first condition, in that it can be hard to identify true preferences without direct evidence. Thus it is one thing to ask people what their preferences are, through a visitor survey, for example, but quite another to offer choices and witness visitor reaction, especially since recreation demand must, to some extent, be supply-led. The second and third conditions have no specific difficulties attached to them, beyond the ability of the manager to define and separate different market segments.

The marketing strategy can, therefore, consist of an undifferentiated policy, concentrating only on factors common to all market segments; a differentiated policy with different marketing plans for different segments; or a concentrated policy, where all marketing effort is placed on specific market segments. Miles and Seabrooke (1977) suggest that concentrated marketing is likely to prove the most applicable to recreation ventures, partly due to resource limitations and partly due to the limited geographical accessibility of any individual site. There are instances, however, when these considerations may be overridden, such as at Alton Towers where the geographical market may consist of the United Kingdom and where resources are sufficient to carry out this level of marketing.

Once the potential market has been fully delineated, attention can be turned to forecasting future demand. Field and MacGregor (1987) suggest that the forecasting techniques available for sites or facilities fall into three categories: trend lines; the Clawson method; and gravity models. Trend lines are often the most straightforward method of forecasting due to the usual availability of attendance data for sites or facilities. In most cases a straight line trend will be appropriate with data corresponding to monthly or quarterly attendances. The periods for which data are collected will depend on the venture, but monthly figures may be necessary to examine seasonal fluctuations, whilst weighted averages may be necessary to establish longer term trends that otherwise might be masked by seasonal fluctuations.

The principal advantage of trend line forecasting is its simplicity and the likelihood that data will be available. It does, however, have some serious drawbacks. These principally concern the implicit assumption that the future will correspond to the past and present and that no new facilities will open to offer competition.

Another approach to forecasting is the Clawson method, developed by Marion Clawson some thirty years ago (Clawson and Knetsch 1971 provide a full description). This method seeks to explain the number of visits to a site or facility by the cost of the visit. This means that visitor numbers may then be forecast according to the entry price and total cost of visit. However, as Field and MacGregor (1987) indicate, whilst this method has become well established in academic literature there is little evidence of its use in practice. This is largely due to the problems inherent in the method, mostly concerning the estimation of costs, the relationship between travel and entry price and the flawed central assumption that visitor rates are determined by cost alone (Field and MacGregor 1987).

A final method of forecasting demand is by the use of a gravity model.

Gravity models are based upon the observed behaviour of urban systems where the amount of interaction between two areas is largely dependent upon their respective size and attractiveness, and the distance between them:

> The proposition is then, quite simply, that the amount of interaction between two areas, A and B, is related directly to the size or attraction of the areas and inversely to the distance separating them (Field and MacGregor 1987, 104).

Once a gravity model has been established for an area the effect of a new development can be tested. Included in the test can be factors such as the likely level of use, traffic generation and effect on other existing facilities. However, in common with other methods of recreation forecasting, estimates can only be made from existing trends and cannot deal with aspects such as latent demand. Equally, the gravity model relies upon an ability to isolate the areas to be studied. This means that it is more suited to modelling the relationship between cities and remote natural areas than it is for competing recreation attractions in densely populated areas.

In conclusion, it can be seen that the use of forecasting in the planned provision of recreation facilities is fraught with many problems. At one level these involve the very nature of recreation provision itself, where supply and demand, or at least consumption, can be inextricably linked. Thus any policy decision to increase supply may also increase future consumption levels. A further issue affecting recreation forecasting is the problem of predicting, or allowing for, future trends, since it is often dangerous to base future forecasts on past trends, or to attempt to predict demand more than five years ahead. However, until much more is known about the complex causal relationships conditioning recreation participation, even the most sophisticated attitudinal surveys are unable to break the constraint of the 'future like the present' assumption.

Because of these problems, it is doubtful to what extent private commercial firms use such forecasting techniques when examining the feasibility of a proposed development. Indeed, the scarce evidence that exists tends to indicate that some very simple criteria are used as a first stage of feasibility, followed by the use of professional judgement, normally more associated with the public sector, in determining whether the venture is likely to succeed. Thus, the Rank Organisation use population, access and competition as their principal criteria for new multi-facility recreation centres, with only those passing the initial feasibility criteria receiving more detailed attention. As Field and MacGregor state:

> Forecasting and planning should be linked but, with unreliable models whose record of accuracy is poor, it is not surprising that this is rarely the case. At the local scale, where the models are more difficult to use and where data limitations may be more severe, there is often recourse to planning based on standards. The limited forecasting that does take place may be used to validate a policy, to choose between plans, or to assess the impact of a proposal. When, as is often the case, supply is constrained, this may not be a problem (1987, 182).

4.4 Financial viability

4.4.1 Introduction

Whatever the motives for deciding to provide a leisure facility, a developer must take account of financial considerations. Of immediate concern will be the cost of land and construction, together with the means and cost of raising the necessary funding for the scheme. Of at least equal importance, however, will be the net cost of operating the facility once it is open and the extent to which the operation of the facility can achieve the operator's aims and objectives. Indeed, there are few aspects of the development and operation of a leisure facility that do not involve financial considerations, regardless of whether the facility is operated by a public authority or a commercial company (see Figure 4.3).

In Figure 4.3, the outline structure of the financial appraisal, management and control of a leisure facility, it can be seen that the basis of the appraisal is the set of aims and objectives held by the developer or operator. The aims of an organisation can be described as broad-based intentions concerning what that organisation wishes to achieve. In the private commercial sector aims are often related directly to financial performance, such as profit maximisation or the achievement of predetermined rates of return on investments. In the public sector they are typically concerned with the maximisation of social welfare or the redistribution of wealth. This is not to deny, however, the non-financial motivations of the private sector (Bevins 1971 and Gratton and Taylor 1987A), or the financial elements of public sector aims, such as the achievement of pre-determined rates of return on money invested (as illustrated by Likierman 1979), or the financial control exercised through local authority budgets. Aims are, thus, long-term statements about the intended end result of particular courses of action or modes of operation.

Objectives are more narrowly defined than aims and relate to short- and medium-term methods of progressing an organisation towards the fulfilment of its aims. Thus, it could be an objective to maximise the use of a leisure facility where the aim is to provide sport for all the members of the community. Equally, an organisation could develop financial objectives, such as income or profitability levels, to achieve both financial and non-financial aims. By being more specific, objectives are more clearly capable of measurement than aims; they are also less likely to vary between different sectors of the economy.

Regardless of sectors in the economy, however, all facility developers and operators will be interested in financial efficiency; the ability to get the most output from the least resources (Stabler 1990 and Gratton and Taylor 1987). This begs the question of what the 'most output' actually represents. Whilst it might be simple to define it as the greatest number of people using a facility, it is more correctly concerned with the provision of the greatest capacity for

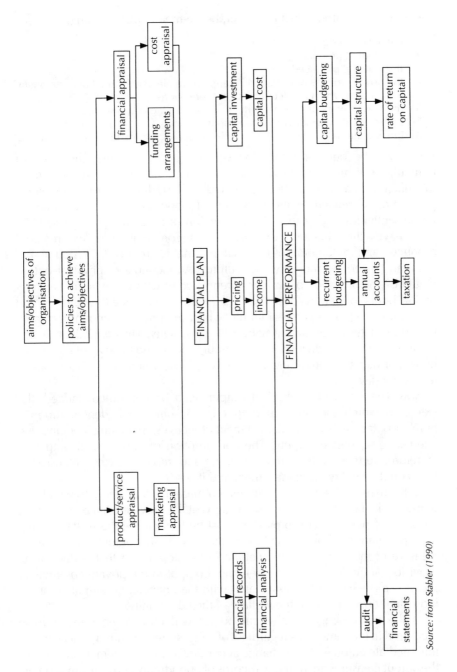

Figure 4.3 Structure of financial apparaisal, management and control

Source: from Stabler (1990)

use. The term 'output' itself has a number of meanings, including:

1. the act of production or manufacture;
2. the amount produced;
3. the material produced;
4. the power, voltage or current delivered by an electrical circuit or component;
5. the point at which an electrical signal is delivered;
6. the power, work or energy produced by an engine or a system;
7. the information produced by a computer; and
8. the operations and devices involved in producing information from a computer.

Whilst all the definitions are in common usage, the first three are of particular relevance to this book. The first, the act of production or manufacture, has already been repeatedly referred to as 'provision', since production is related to the provision of opportunities to participate in leisure activities, rather than the production of an actual good or service. This leaves the definition of output as a combination of the material produced and the amount of that material that is produced. This combination can be termed the 'physical output'. An example of physical output would be the provision of an indoor tennis court with a maximum capacity of 24 hours of tennis per day. The material produced would be the opportunity to play a game of tennis, whilst the physical amount available would be determined by the nature of the activity. Thus we might expect tennis sessions to be divided into one hour blocks, and to be available for 15 hours per day, thus providing an effective physical output of 15 sessions of tennis per day.

However, not all the physical output need be consumed. Indeed, the extent to which the physical output is consumed will depend upon a number of factors, such as management objectives, marketing, demand and alternative sources of supply. The consumption of the physical output is, therefore, termed 'intermediate output' and represents the relationship between the facility, its management and its use.

Lastly, there is the 'final output', which measures the degree to which the output of the facility has achieved the aims of the provider. Since, ultimately, the aims of public sector provision must be to increase the welfare of the local population, it follows that the magnitude of final output will involve a measure of the change in the welfare of the population that is due to the provision of the facility. Conversely, final output in the private commercial sector is more likely to be associated with the financial performance of the individual firm, than with the welfare of the community.

As the name suggests, physical output will remain the same and be measured in the same units regardless of value system. Similarly, measures of intermediate output will not change according to value system. In both cases the output measures are numeric records of opportunity and consumption. As Peacock and Godfrey state, with respect to public museums and galleries:

> On the output side, a gallery or museum presents a range of products in the form of exhibits which satisfy a desire for visual enjoyment or for instruction. The

important characteristic for this range of products is that enjoyment of them cannot be traded. The enjoyment entails consumer participation in the "process of production", as, for example, in the case of a surgical operation ... . As in other forms of consumption, it is not unreasonable to postulate that even those who do not themselves enjoy the products may derive a satisfaction from their enjoyment by others. (1975, 135).

Thus the physical output of leisure facilities is measured as the amount of opportunity available to participate in given leisure activities. Since this opportunity cannot be stored for future consumption, intermediate output is the measure of how much physical output is taken up by consumers. Both these types of output are measured against the facility rather than the aims or objectives of management, meaning that the units of measurement can remain the same in all circumstances.

Since final output is related to the aims and objectives of the provider, measurement must be compatible with those aims and objectives and must be capable of measuring the degree to which the facility provision and management has achieved them. If the primary motivation of the provider is financial the final output would be expressed in financial terms. Equally, if the principal aim of the provider is concerned with social welfare the final output would be expressed in those terms. Thus, unlike physical and intermediate output, the measurement of the final output of a leisure facility will be dependent upon the aims and objectives guiding its operation (see Figure 4.4).

Inevitably, the reasons why the facility is provided will affect what is provided and by whom it is consumed. The nature of the physical and intermediate outputs would, therefore, be dependent upon the nature of the intended final output; they would, however, still be measured in the same terms, regardless of that final output.

Whilst aims and objectives may differ quite radically between providers, all organisations will have financial responsibilities to their employees, clients, shareholders or community (Stabler 1990). Thus, having defined their aims and objectives, developers and providers need to undertake product and financial appraisals in order to develop a financial plan for the proposed facility (see Figure 4.3). These appraisals can be divided into three constituent parts: cost; funding; and pricing. Once these elements have been analysed the developer will be in a position to decide whether or not the proposed scheme is viable and if so, to draw up the resulting financial plan. The remainder of this section will consider cost, funding and pricing, whilst the rest of Figure 4.3. will be dealt with in later sections, when the performance or success of the development is discussed.

4.4.2 Cost appraisal

Cost is incurred in acquiring inputs to develop and operate a leisure facility. Inputs can be defined as the resources used in the development and operation of the facility to supply a set of outputs. The four basic inputs are

	Input		Physical output		Intermediate output		Final output
Public Sector	Land Capital Labour Management	→	15 × 1 hour sessions of tennis	→	Say, 10 games of tennis	→	Increased welfare of community Based on such measures as: a) improved health of society; b) reduction in boredom/ delinquency; c) international sporting success; d) equal opportunity to play tennis.
Private sector	Land Capital Labour Management	→	15 × 1 hour sessions of tennis	→	Say, 10 games of tennis	→	Financial return to individual provider; based on measures such as: a) gross/net profit; b) return on capital; c) cash flow; d) return on turnover.

Figure 4.4 The nature of input and output in the public and private sectors of the economy: the example of a one-court tennis centre

land, labour, capital and management. The first three of these represent the physical resource, whilst the fourth reflects the ability to combine them in a way suitable to achieve the desired output. The nature of the inputs does not vary, regardless of the type of facility or the use to which it is put. Thus, for a leisure facility, land will be required, either as a primary resource (as in the case of a country park), or as a base for developing a built facility. Labour will be required to develop and run the facility. The capital will comprise any built facilities, equipment or service infrastructure. The management input will not only concern the operation of the facility, but will also extend to combining the resources in the development of the facility and altering the input mix to achieve different outputs.

In common parlance the measure of the cost of these inputs is usually represented by the total money expenditure, or price, necessary to acquire them. However, over-emphasising money as the medium of exchange can obscure the true nature of cost; it should be more accurately expressed as the cost of the opportunities foregone (but since money is the medium of exchange even the opportunity cost will be expressed in monetary terms). The notion of 'opportunity cost' can be expressed as a measure of what has to be given up in order to achieve a desired aim (Bannock *et al.* 1978 and Convery 1976). Thus the cost of spending £100 on the membership of a squash club is not the £100, but the foregone opportunities of spending that £100 on other items. Conversely, the price of the squash club membership is £100, since this is the amount of money (the medium of exchange) necessary to acquire it. Whilst this meaning may accurately reflect the nature of cost, it does not lend itself easily to financial analysis. In order to approach this problem the concept of implicit and explicit costs can be considered (Blair and Kenny 1982).

Costs are deemed to be explicit when the item purchased was obtained in the open market. This is because a single purchase does not significantly alter the availability of the item or the price at which it is sold. An example of this might be a sports centre manager purchasing footballs. Anyone can purchase footballs and more are available at the same price. Consequently, the price paid represents the explicit, and opportunity, cost of the item in question.

Implicit costs, conversely, do not involve an actual payment; they consist of the use made of assets already in the possession of the facility or developer, or those obtained at less than the true opportunity cost. An implicit cost arises, for example, where a sports centre is developed by or for a local authority on land already owned by the authority, but where no monetary value is attached to that land when calculating the cost of development. Since the sports centre could be sold or let to another organisation, the implicit opportunity cost of this action is the price that the facility could have commanded in the market. Implicit costs such as this can be considerable and a failure to recognise or account for them could lead to an under-valuation of the total costs of a facility, or its total value on disposal.

Two main categories of explicit cost can be isolated, namely capital and operating costs. Capital expenditure can be defined as that 'which results in the acquisition of an asset' (Kempner 1980, 65). Examples of assets in leisure provision would include inputs such as buildings and fixed equipment. Capital costs do not cease once a development is complete, but will be incurred whenever new equipment is purchased, or whenever the facility is renovated or altered in any way. Operating costs, by definition, must include all expenditure incurred in the operation of a facility. Cleaning, maintenance, energy and staffing would be some of the main costs included in this category.

Since most leisure facilities will last for more than one financial year both capital and operating costs will be incurred over a period of years. Because of this the costs should be treated as flows rather than once-and-for-all payments. The concepts of costs-in-use (Stone 1980) and life-cycle cost (Stonor 1983) have been developed to take account of the time phenomenon. Cost-in-use would include all costs incurred from the date that the development is completed and opened for use, whilst life-cycle costs are the amalgam of costs-in-use and all the initial costs of site acquisition and works, construction, fixtures, fittings and equipment, as well as professional fees.

The list of implicit costs will depend largely upon what has been included in the category of explicit costs. Interest payments and sinking funds will become implicit if they are not fully charged against the facility, as will any central labour or administrative work supplied by local authorities and not charged for. Items such as environmental and social costs should also be considered (Convery 1976), even if their quantification is, at best, subjective.

To conclude, therefore, the term 'cost' is an expression of the opportunities foregone in using any good or service in a given way. The cost need not be met by the user and may not be capable of quantification in any known value system. If the cost of a good or service was met by a third party its price to the user would be nil. If nobody actually met the cost of a good or service it would be deemed to have an implicit cost but no price.

The issue facing a recreational facility developer is, therefore, to translate these definitions of cost into an appraisal of the scheme under consideration. Generally, in the world of speculative property development, the developer will be primarily concerned with the explicit capital cost of the scheme, although recognising that a prospective purchaser is likely to pay more for a building with recurrent cost saving features. The same will probably be true for a local authority commissioning the development of a leisure facility where, since the capital and recurrent budgets are separate, it may be preferable to specify a low capital cost building, regardless of the future cost implications, simply to be allowed to go ahead with the development. However, there are sectors of the leisure market where recurrent costs are considered at the outset:

A discotheque is built to be completely refurbished every four or five years, with the financial modelling and return on investment calculations designed to show this. In discotheques, bingo halls and even cinemas, refurbishment and main-tenance are not seen as optional extras, only to be afforded if there are sufficient residual funds. The private sector treats refurbishment and maintenance as unavoidable business costs and they are budgeted for in the same way as staff, rates and insurance (Lynch 1989, 85–86).

As Lynch (1989) implies, the problem of cost appraisal is that it is specifically related to the type of development under consideration, with only those experienced in particular types of facility development having access to comprehensive sets of data. However, it is equally apparent that only recently has the issue of cost analysis in recreation development been considered. Although some authors suggest that the awareness of the importance of leisure facility costs stemmed from the growing ideological attack on state provided goods and services (Hatry 1983 and Collins 1986), empirical evidence indicates that the awareness of the life-cycle costs of leisure facilities occurred at about the same time as it did for other buildings and capital assets.

A body of information that does exist is that associated with cost-efficiency. However, the primary motivation behind much of this work is actually cost-cutting, or cost-saving (Ontario Ministry of Tourism and Recreation 1983 and Sports Council 1976). The Sports Council study of direct and joint sports provision (Coopers & Lybrand Associates Ltd 1981) considered cost efficiency in a number of different types of leisure facility. Whilst ascertaining that jointly provided centres (those provided partly by the local education authority) tended to have higher cost-efficiency ratios than separately provided centres (those provided solely by the recreation department of the local authority), the issue of total cost was not addressed. Furthermore, since the cost-efficiency ratio was defined as usage divided by operating costs the subject of capital cost was not considered in this context, but only as a separate issue.

A study of the cost-effectiveness of developing and operating leisure facilities was undertaken by the Audit Inspectorate in the early 1980s (Department of the Environment, Audit Inspectorate 1983). Whilst the objectives of the study were to conduct a comparative study of '... all aspects of a centre's development from its conception to its opening to the public, and ... an investigation of the operation and management of leisure centres' (Department of the Environment, Audit Inspectorate 1983, 1), little emphasis was placed on financial analysis.

Following recognition of the need to assess both initial and recurrent costs of sports buildings (Birch 1971) and 'a constant stream of enquiries' concerning the magnitude of sports centre running costs (Sports Council 1975), the Sports Council developed a framework for establishing the running costs at any individual centre (Sports Council 1975). However, it was little more than a check list of recurrent cost items, with no comparative information to help centre managers assess their financial performance. This report was followed by a booklet detailing methods of reducing sports centre deficits (Sports Council 1976), again without the necessary information for even the most rudimentary financial analysis.

By 1977 the Sports Council had once again reached the conclusion that there was a need to consider both initial and recurrent costs before a complete assessment of the financial implications of leisure provision could be made (Sports Council 1977). Veal (1979) stressed, in a report on low-cost sports centres, that 'low-cost' in initial capital terms did not necessarily mean low operating or maintenance costs. This was concurred with in a later study of low capital cost leisure provision, where the Sports Council commented that:

> Staffing, running and maintenance costs over the life of the building may erode the low capital cost advantage, showing that the initial low cost concept is merely a palliative providing temporary relief in the short term to our cost problems. Further monitoring and research into the total costs-in-use using modern appraisal methods needs to be completed before a final cost judgement can be made (Sports Council 1980, 7).

It remains, however, that such an appraisal has not, seemingly, been applied to leisure facilities, even with the advent of a design guide for various sports facilities backed by the Sports Council (John and Heard 1981, 1981A, 1981B and 1981C). Indeed, very few references are made to the cost of leisure facilities at all in the guide, with the main references concerning broad elemental breakdowns of capital costs for ice rinks, swimming pools and sports halls. The only mention of recurrent or future costs are the subjective statements that there is value to be gained by taking future costs into account (John and Heard 1981), and that future maintenance costs will not be significant when compared to capital, staffing and operation costs (John and Heard 1981A).

A similar lack of emphasis on the analysis of total life cycle costs is evident when assessing the work undertaken by other government agencies such as the English Tourist Board and the Countryside Commission. The significance of this oversight becomes apparent when Arthur Young McClelland Moores and Co. state, with respect to research on the economic efficiency of national park authorities, that:

> ... major projects involving large capital sums or the undertaking of liabilities have sometimes gone ahead without full justification in terms of need, appropriateness or the subsequent consequence for continuing maintenance costs. (1984, 27).

The authors go on to suggest the need for selective performance assessment for national park authorities, including the capital and revenue financing implications of developing and operating recreational facilities. In doing so, however, they note that the Countryside Commission must put more weight on these requirements if there is to be adequate support and feedback to the national park authorities.

The English Tourist Board has a longer history of considering the financial implications of leisure provision. An early example of this, concerning static holiday caravans and chalets (English Tourist Board 1973), considered initial development costs and charging structures, but did not account for administration and maintenance costs. Along with grant aid, the English Tourist Board has produced information to help individuals considering investing in a tourist project. Whilst much of this information was available in consultation with Board officers, the Board has also published a series of Development Guides, which are updated periodically. Although some of the guides are very general (English Tourist Board 1985, for example), others do address the issue of initial and recurrent costs. An example of this concerns the management of guest houses (English Tourist Board 1984) where a list of running cost items is given, together with the statement that running costs should account for about 40 per cent of gross income.

The Scottish Sports Council produced a major study of sports facilities in 1979, the basis of which was an attempt to link the economic implications of facility provision with the 'user patterns produced by this provision' (Scottish Sports Council 1979, 2). Whilst recognising that a systematic determination of cost structures is a necessary prerequisite to determining cost-efficiency, the study concentrated on specific leisure facilities rather than developing a standardised methodology applicable to any individual facility (Casey 1980).

However, the study did consider total costs, defined as operating and capital expenditure, to produce some interesting results, for the year 1975/76. The operating cost per visitor was found to vary between different types of facility, but to be relatively constant at 50 to 52 pence per visitor for swimming pools and sports centres with swimming pools. When considering the breakdown of operating costs the Scottish Sports Council state that:

> The most obvious conclusion to be drawn from the data ... is that there is a remarkable similarity between the structure of costs at many of the centres in the study. (1979, 43).

It should be recognised, however, that in amalgamating operating and capital costs, only those incurred in the study period were recorded. Thus, the capital cost recorded was solely that incurred in the one financial year under consideration. This means that although proportions are shown for capital and operating costs, the actual proportions have little bearing on the life-cycle cost, because:

> ... the annual capital charge is generally relatively fixed while the operating cost rises annually according to the costs of factors of production such as wages and fuel (Scottish Sports Council 1979, 45–46).

In contrast to the Scottish Sports Council's (1979) findings, work on other building types has produced differences in the cost structures of similar facilities. The work on offices (Purkis *et al.* 1977 and Beeston 1983) and hospitals (Michel 1979 and Akehurst and Blackburn 1979) suggests that factors other than the actual buildings can be significant in determining the magnitude of total life-cycle cost. Purkis *et al.* (1977) suggest local factors or variants not analysed in their study may account for cost variations and Beeston (1983) suggests that no two buildings are exactly alike and will not, therefore, have similar cost structures. With respect to hospitals, Michel (1979) suggests that differences in cost occurred due to different inter-pretations of the standard of care supplied by each hospital; in other words, that the management of the hospital is a significant determinant of total building cost. Whilst apparently being at odds with these studies, it may be that the management of the sports facilities (Scottish Sports Council 1979) was sufficiently homogeneous that differences in cost were not detectable, or that the study had insufficient data.

In their study of schools in Nottinghamshire, Hoar and Swain (1977) found that recurrent costs could amount to 20 per cent of the total capital costs for each year of operation. This means that, over an economic life of 30 years, the total cost of a school will amount to seven times the initial capital cost, assuming no major renovations are undertaken in that period.

Similarly, Meggitt (1980) describes the modelling of the cost of new cities overseas by illustrating how the relative magnitude of life-cycle cost elements vary from one type of facility to another. The range of facilities varies from hospitals, where operation and maintenance can be responsible for 80 per cent of life-cycle costs, to steel pipelines, where the same category accounts for a mere four per cent. Whilst no figures are given for any type of leisure facility, some are available for schools, which do share many of the same characteristics. These suggest that less than 20 per cent of life-cycle cost can be attributed to capital costs, about 60 per cent to operation and maintenance and 20 per cent for renovation and renewal.

Recent figures produced by Adie (1985) for a proposed leisure pool in Blackburn suggest that the life-cycle cost structures of leisure facilities are more closely associated with those of hospitals than those of schools. According to Adie, both capital and recurrent costs had been calculated by Blackburn Borough Council with the result that over a 60 year life:

> The balance will be: capital £2.15 million; upkeep £34.8 million at today's prices. The income from entrance fees ... will reduce this by 33 per cent, giving a net life-cycle loss of £21.5 million – 10 times the original cost. (1985, 67).

Ignoring the income element, since it was not included in Meggitt's calculations, the figures indicate that over 90 per cent of the total life-cycle cost is attributable to operation and maintenance.

From this review it is clear that those wishing to assess the financial viability of a leisure development are faced with a paucity of information on cost, regardless of whether their interest is primarily associated with the capital cost, as in the case of a speculative developer, or total life-cycle costs.

With so little information available on leisure facility costs, the Sports

Council undertook the development of the Standardised Approach to Sports Halls (SASH) initiative, in conjunction with Bovis Construction Ltd (Sports Council 1985). In commissioning the design of a sports hall that represented 'value for money' in life-cycle, as opposed to solely initial, cost terms (Sports Council 1982), the emphasis was on the quality of initial construction, furnishing and fittings, in the belief that this would lead to low maintenance costs in later years and to a superior quality of facility throughout its life.

Whilst there have been claims that the design selected is not the most appropriate for a low cost facility, and whilst rather fewer authorities have ordered one than was initially hoped, the SASH sports hall does represent a useful case study in cost analysis. The development of the SASH scheme was announced in September 1982 in a 'move to speed up provision of indoor sports facilities in England' (Sports Council 1982). It was based on the standardised design of a low cost sports hall 'which is cost-effective in terms of construction, maintenance, repair and management charges' (Sports Council 1982). By grant-aiding 50 per cent of the initial development cost for the first nine SASH centres (one in each Sports Council region) it was hoped that the impetus would be provided for the development of a total of 60 centres by 1987. In the event, fewer than half of these have been built. The basic SASH facility consists of a 4-badminton court sports hall, fitness room, social area and ancillary facilities, with possible additions including squash courts, a swimming pool and an additional sports hall. The majority of those built have been of the basic format.

The initial capital cost of a SASH centre was calculated for the Sports Council and included in the design guide produced in 1985. The figures are given in Table 4.4.

Table 4.4 Standard SASH cost plan (at 3rd quarter 1985)

Element	Total cost (£)	% Cost of building
1. Substructure	38 000	8.43
2. Superstructure	182 000	40.35
3. Internal finishes	43 000	9.53
4. Fittings	44 000	9.76
5. Services	103 000	22.84
6. Preliminaries	41 000	9.09
Net Cost of Building	451 000	100.00
7. External works	61 000	
Total Construction Cost	£512 000	

Source: Sports Council 1985A

From the capital cost plan it can be seen that the majority, some 63 per cent, of initial cost is comprised of the superstructure and services, with the remaining cost elements contributing about 10 per cent each.

The recurrent costs likely to be incurred by the SASH facilities were more difficult to ascertain, with no data available from the Sports Council or Bovis Construction, the designers and developers. This therefore required derivation of the likely costs from a range of sources, but based upon the design guide prepared by Bovis Construction. The recurrent cost estimates are shown in Table 4.5, with a complete explanation of their calculation to be found in Ravenscroft (1988).

Table 4.5 Summary of total recurrent costs for SASH facilities
(prices as at 1985)

Element	Annual Cost £	Periodic Cost £	Repair Cost £
Substructure	–	–	-
Superstructure	25 577	18 247	1 008
Internal finishes	151 122	39 179	–
Fittings	68 601	75 229	–
Services	137 157	88 655	–
External works	21 952	10 258	–
Repairs to damage	–	–	–
General cleaning	20 563	–	–
Staff	1 332 827	–	–
Admin/operation	450 809	–	–
Implicit costs	970 954	–	–
	3 179 572	231 567	36 436
Total	3 447 575		

As can be seen, the most significant elements of recurrent cost are associated with staffing and implicit costs. Whilst it is recognised that staff costs will be a major feature of the overall cost profile, the implicit costs are artificially high to cover the cost of rent, rates and interest payments, all of which could be identified separately. What is abundantly clear, however, is that some of the major costs are not related to the facility as much as to the ownership and use of the facility; a point that would be more graphically illustrated if the implicit costs were reallocated, largely to administration and operation.

When the two sets of cost figures are combined, as in Table 4.6, it is clear that the annual cost category is the most significant in determining the life-cycle cost, contributing over 80 per cent of the total. When the periodic and

Table 4.6 Elemental life-cycle cost plan for SASH facilities (all prices as at 3rd quarter 1985)

Element	Initial construction costs (£)	Annual costs (£)	Periodic costs (£)	Repair costs (£)	Total (£)
Substructure	41 800	–	–	–	41 800
Superstructure	200 200	25 577	18 247	1 008	245 032
Internal finishes	47 300	151 122	39 178	–	237 600
Fittings	48 400	68 601	75 229	–	192 230
Services	113 300	137 157	88 655	–	339 112
External works	61 000	21 952	10 258	–	93 210
Repairs to accidental damage	–	–	–	35 428	35 428
General cleaning	–	20 563	–	–	20 563
Staff	–	1 332 827	–	–	1 332 827
Administration and operation	–	450 809	–	–	450 809
Implicit costs	–	970 954	–	–	970 954
Total	512 000	3 179 572	231 567	36 436	3 959 575
Percentages of total life-cycle cost					
Substructure	1.0				1.0
Superstructure	5.0	0.6	0.5		6.1
Internal finishes	1.2	3.8	1.0		6.0
Fittings	1.2	1.7	1.9		4.8
Services	2.9	3.5	2.2		8.6
External works	1.5	0.5	0.2		2.2
Repairs to accidental damage				0.9	2.2
General cleaning		0.5			0.5
Staff		33.7			33.7
Administration and operation		11.4			11.4
Implicit costs		24.5			24.5
Total	12.8	80.9	5.8	0.9	100.0

Note: Preliminaries have been apportioned to the five elements constituting the net cost of building. Figures may not sum to 100 due to rounding.

repair cost categories are included, recurrent costs contribute over 85 per cent of the total life-cycle cost of a SASH facility.

What is also abundantly clear is that a building that cost £500000 to build is likely to have cost the owner and operator over eight times as much by the end of a planned 30 year life. Furthermore, the proportion of this figure attributable to capital inputs fails to reach 20 per cent of the total, even once all the capital items in the recurrent cost category have been reallocated, as in Table 4.7.

Table 4.7 SASH life-cycle cost profile 4

Element	Cost	%
Capital	707 185	18
Staff	1 332 827	35
Maint/repair	207 473	5
Admin/operation	741 126	20
Sinking fund	307 726	8
Interest on capital	533 856	14
Total	3 830 193	100

Finally, the total life-cycle cost plan can be recalculated to provide an annual cost model, as in Figure 4.5. This suggests that a SASH facility should have cost, at 1985 figures, in the region of £195 000 per annum to develop and operate. In terms of operation expenditure alone, the cost falls to approximately £125 000 at 1985 prices. This annual cost is, by its nature, an approximation since some of the costs, particularly that for renovation, do not occur uniformly every year, but are incurred in larger 'lumps' periodically.

The final factor to consider at this stage is income to the facility operator. All the figures given for the SASH facilities have been gross, with no allowance for any income that may be earned. The reason for this is that the income derived from the facility will be a function of the marketing and pricing policies. As such, income is actually part of the output of the facility and should, therefore only be related to the costs at the stage of monitoring and performance evaluation.

Whilst this might be theoretically valid, it must be recognised that the aims and objectives set for the facility may link cost and income, or that the magnitude of one may be dependent upon the other. This would occur in the case of a commercial operator where a specified rate of return on capital or turnover might be expected, or in the public and voluntary sectors, where income might be expected to cover costs. This may lead, in all cases, to cost being measured in a different way, such as by ignoring all implicit costs, or by measuring cost against revenue rather than against cost standards.

In conclusion, this section has shown that, whilst there are few available data to aid in establishing the likely cost of a proposed facility, it is possible

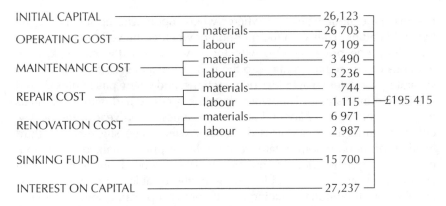

INITIAL CAPITAL —————————————————— 26,123 ⌐

OPERATING COST ————┌— materials————— 26 703 —
　　　　　　　　　　　└— labour ————— 79 109 —

MAINTENANCE COST ———┌— materials————— 3 490 —
　　　　　　　　　　　└— labour ————— 5 236 —

REPAIR COST ————————┌— materials————— 744 —
　　　　　　　　　　　└— labour ————— 1 115 —├—£195 415

RENOVATION COST —————┌— materials————— 6 971 —
　　　　　　　　　　　└— labour ————— 2 987 —

SINKING FUND ————————————————— 15 700 —

INTEREST ON CAPITAL —————————————— 27,237 ⌐

NOTES

1. Assume a 30-year life of facility
2. All figures are calculated at 1985 prices
3. All figures are discounted at 3 per cent per annum, with no allowance for tax or inflation.

Figure 4.5   A normative annual cost model for SASH facilities

to produce comprehensive estimates of the likely cost structure. The analysis used could equally be applied to an outdoor facility, or to one where there is no appreciable development or operating cost. Finally, this is, of course, merely the first stage in determining financial viability, let alone financial performance. The likely, or normative cost structure of the SASH facility will, therefore, be returned to in the evaluation, to be discussed in Section 4.6.

4.4.3 Pricing

When considering pricing in the context of recreation development it is important to distinguish between the process of tendering for the development contract and analysing the relationship between the eventual price to consumers and the level of facility use. Apart from the cost of the inputs, the former is more concerned with the extent to which the developer wishes to win the contract. The latter, in contrast, is one of the most crucial aspects of the development process, in determining the feasibility of the proposed venture. Directly, this is because of the relationship between price and the number of users, but as Stabler points out, its effect on the financial state of the business can be just as significant:

> ... the importance of establishing the appropriate price in leisure provision lies on the impact it has on the cash flow as a result of changes in revenue. In turn this holds implications for the long-term value of the enterprise. (1990, 118).

When seeking to establish a possible pricing policy and structure, the developer is faced with a range of options. Bovaird *et al.* (1984) group these options under four headings: where; when; upon whom; and how. The 'where' heading deals with the question of whether the venture is designed for a 'pay one price' system, with the pay booth in the car park, train terminus or the actual entrance to the park, or whether visitors should pay separately for different activities or attractions, and where pay booths should be situated in this case. The 'when' heading is concerned with differentials in pricing that may be applied at different times of the year, week or day, either to stimulate or suppress demand. As the name suggests, the third heading of 'upon whom' relates to the relative prices charged to different people. Thus, it is common practice to offer reduced prices for certain groups, such as elderly people, children, the disabled and large parties. Equally, the discretionary pricing to different groups can also be related to the time-differentiated prices, by only allowing, for example, the reduced rates to apply at off-peak times. Finally, under the 'how' heading it is important to consider such aspects as the different technical means of collecting money and distributing tickets, the case for offering memberships or season tickets, and whether there is any advantage in issuing joint tickets with complementary facilities in the neighbourhood.

In addition to determining the pricing structure for the proposed facility, the developer should also consider the actual price to be charged. This will be dependent upon a range of variables. The most important of these are the socio-economic characteristics of the target market, the marketing process, the type and quality of experience to be created, the competition expected and the management of the venture. However, as Bovaird *et al.* (1984) discovered, the seemingly fundamental importance of pricing does not mean that it has been highly researched, or that the relationship between prices and visitor numbers is well known by operators or researchers. They conclude that:

> ... little guidance is available to site owners and managers on the actual relationship between levels of charges and levels of use or on the most appropriate pricing strategy to adopt in order to meet local needs and circumstances. (Bovaird *et al.* 1984, 1).

The basis of this relationship is known as the elasticity of demand. This is the degree to which the consumption of a good or service is affected by changes in the price at which the good or service is offered. Thus, at one end of the scale is elastic demand, where consumption decreases more than pro-portionately to price increases, whilst at the other end is inelastic demand, where the opposite reaction to price increases is encountered. In the middle, displaying proportionate responses to price changes, is unitary elasticity of demand.

By establishing the elasticity of demand for a proposed facility, or at least class of facility, it should be possible to determine how visitors will react to different prices, so enabling the manager to set the most appropriate price to achieve the aims and objectives set for that facility. Using records of past visitor numbers and price changes at a range of sites operated by the National Trust and the Department of the Environment (the latter now being run by English Heritage), Bovaird *et al.* (1984) found a fairly inelastic demand for most DoE sites, with visitor numbers being less than proportionately affected by price increases. In terms of pricing policy, therefore, Bovaird *et al.* concluded that:

> These results indicate that there was, in general, substantial scope for raising revenue through price increases at DoE sites during the study period. However, this also means that using admission prices as a tool for restraining visitors at heavily used sites would often necessitate either very large overall price increases or special pricing arrangements for peak periods only. (1984, 150–151).

Results from the National Trust sites tended to indicate higher elasticities, particularly at the more popular venues. This showed that, in general, demand for visits to the National Trust sites are more price-sensitive than those of the DoE but, equally, that visitor restraint through price increase would be easier to attain.

Beyond visitor restraint, the elasticity of demand is also important in determining profitability, since the more inelastic the demand for a site or venture the easier it is to increase profitability. This is because as the admission price rises the total revenue to the venture also rises since the fall in visitor numbers is less than proportionate to the price rise; this is further compounded by a marginal decrease in costs as visitor numbers fall. Thus it is to be expected that, in the private commercial sector at least, the providers of recreation experiences and opportunities will attempt to raise prices until demand becomes price elastic (Bovaird *et al.* 1984).

Whilst agreeing with this theory of private sector pricing, Gratton and Taylor (1985) state that available evidence suggests that firms do not conform and, in particular, that they tend to base their pricing decisions on cost alone. This is witnessed in descriptions of pricing policies such as 'full-cost' pricing, or 'cost-plus' pricing, for example. However, it is difficult to know whether demand is intentionally ignored when prices are being set, or whether it is that demand forecasting is felt to be too unreliable:

> ... the emerging picture is that in practice private sector pricing does not conform to the strict profit-maximisation objective. This may mean that profit-maximisation is not the only objective, ... or simply that the information on marginal values necessary to pursue this objective is not obtainable.
>
> (Gratton and Taylor 1985, 159).

A similar picture is found in the public sector where, once again, pricing is more usually based on costs than on the relationship between demand and cost. However, rather than the full cost philosophy of the private sector, those in the public sector are more inclined towards concepts such as marginal cost pricing. At the theoretical level these sorts of policies deny the public the advantage of any consumer surplus derived from the willingness of some consumers to pay more than they are actually charged (Gratton and

Taylor 1985). Equally, at the practically applicable level problems also exist since the marginal private cost of provision is rarely calculable, whilst any attendant social costs and benefits remain unknown (Gratton and Taylor 1985).

Thus whilst Bovaird *et al.* (1984) state that demand analysis is as useful for recreation facilities as for other goods and services, Gratton and Taylor (1985) have found little evidence of its application, in either the public or the private sectors of the economy. Indeed, regardless of the encouragement to relate prices to consumers, if not costs, by the Audit Commission (1983), the commonest method of pricing appears to be the comparative method – charging the same as the competition (Department of the Environment 1981 and Department of the Environment, Audit Inspectorate 1983). Whilst this has the advantage of being both convenient and non-controversial, it neither reflects the strength of demand nor the cost of supply (Gratton and Taylor 1985).

In conclusion, therefore, when attempting to assess the likely admission price for a proposed recreation venture, the developer is in somewhat of a vacuum where the theory may exist but the practice is not evident. For those in the public sector the pricing decision may be sufficiently political to render the economic theory irrelevant, even with the onset of compulsory competitive tendering. Equally, in the voluntary sector it may be enough to base prices on the explicit costs borne by the club or organisation. However, in the private commercial and quasi-commercial sectors a failure to under- stand the relationship between demand and cost may lead to the venture failing to achieve its full financial potential or, more fundamentally, failing to survive at all. Futhermore, in all sectors of the economy it is important to recognise the role of pricing in visitor control. As Bovaird *et al.* (1984) have pointed out, marginal increases in the admission price of ventures with inelastic demand curves will not restrain visitor numbers. However, recog- nition of the role of pricing, particularly when allied to marketing, can provide management with the ability to manage demand. This has been the case at the National Motor Museum at Beaulieu where Montagu Ventures Ltd, the operators, have been able to maintain their optimum visitor numbers over recent years through a constant visitor monitoring programme that has allowed them to match prices to the level of consumption that they deem appropriate for the site. However, without this time-series of informa- tion, the developer is really left with comparative evidence as the most likely way of establishing an admission price for a proposed facility.

4.4.4 Evaluation of a proposal

By far the most common form of evaluation in the development of leisure facilities is that undertaken by the potential operator. This is simply because most leisure developments are built for clients, with finance raised by the client rather than the speculative developer. For this reason the evaluation is likely to take the form of a two-part appraisal featuring, on the one hand, the cost of development and operation and on the other, the extent to which the facility can meet the needs of those whom it was designed to serve. In the commercial sector these two parts could be synonymous, at least to the extent that the primary aim for providing the facility is to generate profits for the operator. Conversely, in the public sector the parts might be quite separate, with the pre-development analysis concentrating on financial efficiency criteria and the operational monitoring being based on the effectiveness of service delivery. However, developments in both the public and private sectors can be split into their pre- and post-development phases, with the former concentrating on financial analysis and the latter on the monitoring of operational performance. The evaluatory role will, therefore, be discussed here, with the monitoring role dealt with in Section 4.6.

In first considering the financial evaluation of the development, the traditional method in the private sector has been through the use of investment appraisal techniques. In some cases, where the initial cost is so great that any subsequent costs are insignificant, or when the life of the asset is such that the prospect of repair costs will result in replacement, there may be a case for equating investment appraisal with initial capital costs only. This view is enhanced by the strong link between investment appraisal and capital budgeting, where the technique is concerned with which projects should be undertaken, based on return against capital cost, how they should be financed and the total amount of capital expenditure to be incurred (Bromwich 1979). However, leisure facilities tend to incur significant costs throughout their lives, suggesting that investment appraisal in these cases should not be limited to initial costs, but should consider all future costs as well.

There is no definitive method of investment appraisal, but rather, a set of methods that has been devised for different functions and requirements (Lumby 1984 and Butler and Richmond 1990). A simple form of appraisal is the payback method. The philosophy of this method is to calculate the length of time taken for the incremental benefits from a project to 'pay back' the initial capital invested. In comparative terms, the project with the shortest 'payback' would be accepted. Whilst it is simple and relatively easy to predict, it is really limited to circumstances where a specific output is required in the short-term. Under these conditions the most favourable project will be the one that provides the given output (financial or otherwise) and covers its cost in the shortest time.

Another appraisal technique commonly used is the 'accounting rate of return' or 'return on capital employed' method. It is calculated as the ratio of the accounting profit generated by an investment project to the required capital outlay, expressed as a percentage. The criteria for decision making using this method can either be on an 'accept/reject' basis where projects must meet a minimum, or target rate of return, or the 'best' rate of return can be selected. The main characteristic of this method is the strong reliance on financial return compared with initial cost, thus disregarding non-financial returns and the actual nature of the output.

Because of the long term nature of most recreation facilities, the future flow of costs and benefits assumes considerable importance. This has led to the development of investment/consumption decision models capable of taking future income and cost flows into account (Lumby 1984). In these models it is assumed that the decision-maker wishes to be able to appraise all costs, whenever incurred, at a single point in time, such as the present. This requires the use of a discounted cash flow, whereby all cash flows are discounted by a known comparability factor, such as the prevailing or long term rate of interest, so that they all relate to a single period of time:

> Discounting is essential to the comparison of benefits and costs. Its function is to convert future benefits and costs of a project into present value ... at the time of the decision. ... The sum of the discounted benefits less the sum of the discounted costs over the life of the project measures the economic value of recreation investments. (Walsh 1986, 566).

Once discounted to a common base, appraisal can take a number of forms. Since it has been suggested that the performance of investments in real property and built facilities should be judged by the rental income generated and by any capital appreciation (Hetherington 1984), one method of measurement is the Internal Rate of Return. This method is used to establish the rate of interest which equates the discounted value of a project's cash flows to the discounted value of its investment outlay (Bromwich 1979). By using this method the relative performance of different investments can be compared according to their various rates of return. Whilst this may be of importance to investors, it does assume that all receipts are capable of measurement in similar terms and that the most important investment decision is the return on capital invested.

Another appraisal method incorporating a discounted cash flow is the life-cycle costing method. It is defined as:

> ... a method of economic evaluation of alternatives which considers all relevant costs (and benefits) associated with each alternative activity or project over its life. (Ruegg et al. 1978, 2–3).

Since this method involves the consideration of costs over many years, it too must employ a method of presenting current and future costs in equivalent terms (Flanagan and Norman 1989). Such a method is the Net Present Value concept of discounting. This method is used to calculate the present value of a project's cash flows using a pre-determined rate of interest. The rationale is that all receipts and outlays are discounted to a single monetary figure which is either positive, if the project is financially viable, or negative if it is not.

Whilst the financial basis of the net present value concept is suitable for recreation developments in the commercial sector of the economy, its inability to equate financial and non-financial inputs and outputs renders it ineffective in the public and non-commercial sectors of the economy. As Walsh states:

... in private business, society relies on profits and competition to furnish the necessary incentives and discipline and to provide a feedback on the quality of decisions. While this self-regulatory mechanism is basically sound in the private sector, it is virtually non-existent in the public sector. In government, we must find another tool for making choices which resource scarcity forces upon us. (1986, 556).

One of the first and still most widely espoused, if not used, means of dealing with the non-financial nature of some public recreation provision is the concept of cost benefit analysis. Essentially, cost benefit analysis mirrors the commercial sector in seeking to establish whether the benefits of a proposed course of action outweigh its costs. The principal differences lie in the nature of the potential recipients of the benefits and the rigour with which the analysis is conducted:

> Instead of asking whether the owners of the enterprise will become better off by the firm's engaging in one activity rather than another, the economist asks whether society as a whole will become better off by undertaking this project than by not undertaking it, or by undertaking instead any of a number of alternative projects. (Mishan 1976, 13).

Thus the underlying philosophy of cost benefit analysis is the replacement of a narrow financially-orientated price system with one that explicitly takes account of both financial and non-financial costs and benefits. It is usually expressed as a ratio of the present value of net benefits (the consumer surplus) divided by the present value of the opportunity cost of provision. Where the ratio is greater than one the proposed development is socially efficient; less than one and it is socially inefficient; and a ratio of precisely one will leave society no better or worse off.

Good examples of the application of cost benefit analysis can be found in Seeley (1973), where the technique is applied to a range of recreation facilities in Nottingham, including a park, golf course, allotments and a canal. In recognising the social benefits that accrue from the facilities but are either unpriced, as in entrance to the park, or go unaccounted for in determining the value of the facilities, such as enhanced property values or beneficial effects for passers-by, Seeley shows that, in the case of the park, an annual operating deficit of £20 000 could be converted into a net social gain of over £70 000 per annum, indicating that 'Wollaton Park does meet a great social need and is of considerable benefit to the community' (Seeley 1973, 67).

More recently, cost benefit analysis has been applied to the canals operated by British Waterways in an attempt to calculate their worth, in terms of consumer surplus, to society (Miller 1991). Amongst the unpriced benefits considered by British Waterways were contributions to the balance of payments from foreigners visiting Britain and British residents remaining in Britain solely to participate in a canal holiday; a land drainage function that currently goes uncharged; social benefits to users of the towpaths, where no user fees are levied; and the intrinsic benefits of knowing that the canals are available for use now and in the future. The reason for attempting to evaluate the consumer surplus of the canal network was primarily one of justification:

> These unpriced benefits form a bridge between actual revenue derived from charged activities and the cost of maintaining the waterway system. They therefore provide a justification for the continuation of public funding of the waterways, provided their value in economic terms is more than the grant-in-aid. Consequently, for advocacy reasons, there is a need to develop methods for quantifying these benefits. (Miller 1991).

Using a mixture of surveys and cost estimates, British Waterways were able to estimate that an annual contribution of some £16m was made to the balance of payments, that alternative forms of land drainage would cost at least £20m per annum and that the consumer surplus for unpriced informal recreation, using the individual travel cost method, amounted to £0.55 per visit, or some £63m per annum for all visitors. Excluding the value of any intrinsic benefits, therefore, British Waterways were able to estimate that the value of the canals, at 1989 prices, was at least £100m per annum, against government grant-in-aid of just £45m per annum, thus providing important evidence in attempting to maintain or improve future government funding levels.

Whilst recognising that a technique such as cost benefit analysis can provide surrogates of value when the market mechanism has no means of doing so, the outcome of the technique is wholly dependent on the method of calculation used. The most commonly used surrogate is the amount that people are willing to pay for entry to the site, were a charge to be levied. A number of methods have been devised to establish this 'willingness to pay', including the analysis of price changes, consumer surveys and the travel cost, or Clawson technique. In examining the first of these, Bovaird (1991) warns that apparent responses to price changes at sites where prices are charged can be misleading since many other factors may be changing at the same time. Whilst not so subject to this problem, consumer surveys can be heavily biased by respondents saying what they believe will be in their best interests, rather than what they actually think. Because of this the most popular surrogate, as used by Seeley (1973) amongst others, remains the amount that people have spent in travelling to the site. However, in using this method, it must be recognised that the results are subject to some questionable assumptions, such as the notion that travel costs will equate with the entry price that people are prepared to pay and that the sole purpose of the journey to the park was to participate in outdoor recreation (Miller 1991). Furthermore, even the most ardent supporters of cost benefit analysis are forced to admit that there are numerous difficulties involved and that some benefits cannot be evaluated satisfactorily in monetary terms. Whilst recognising these deficiencies, Seeley does suggest that:

> An important advantage of cost-benefit study is that it compels those responsible to quantify costs and benefits as far as possible, rather than resting content with vague qualitative judgements or personal hunches. Furthermore, quantification and

evaluation of benefits – however rough – does give some indication of the charges which consumers are willing to pay (1973, 55–6).

However, even this recommendation cannot justify the result of a cost benefit analysis constituting a prescription for society (Mishan 1976). Since it is primarily replacing a financial price system with one that can take account of a greater range of factors, as long as they can eventually be reduced to financial prices, cost benefit analysis suffers from the same limitations as any other price system. These are, namely, that the analysis cannot distinguish where, and upon whom, the benefits are bestowed, nor the relative number of gainers and losers resulting from the development. The concept is, therefore, probably of most use for evaluating minor developments and additions to facilities, where the range of costs and recipients of benefits are limited and definable (as, indeed, is the case in Seeley's Nottingham examples).

In attempts to account more specifically for the non-financial benefits accruing from public sector provision, other evaluation models have been developed. Of these, the Goals Achievements Matrix and the Planning Balance Sheet have commanded most attention. In applying the former, the authority or operator attempts to match the goals established for a facility or development with the extent to which they have been realised. The strength of this approach lies in the need to set explicit goals and quantifiable measures of achievement, which could include references to financial cost. In adopting this approach, however, the class of benefits is restricted to formally identified goals whilst the ability to use comparative data is hindered by the specific nature of each individual matrix.

The Planning Balance Sheet represents an attempt to overcome the shortcomings of the Goals Achievements Matrix by looking much more widely at the costs and potential benefits of a scheme. In effect, therefore, it is much closer to the philosophy of the cost benefit approach, although using a wider definition of benefit and, through the use of a balance sheet, making an attempt to combine financial and non-financial costs and benefits in the same analysis. The result of this tends to be, however, a qualitative rather than quantitative overview of the scheme under consideration. This really precludes its use in any comparative capacity since any judgement of performance or choice between schemes would also have to be based on qualitative criteria rather than quantitative evaluation.

In recognising the inherent weaknesses of methods which do not seek to quantify all costs and benefits, or which seek to reduce them all to the same medium of exchange, Ravenscroft (1988) extended the use of the life-cycle costing technique to take account of differing operational objectives and outputs in the provision of sports facilities. Starting from the basis that the inputs to a recreation facility comprise either initial capital or recurrent cost items, it is assumed that the initial capital is dictated by the physical structure of the facility, which in turn determines the maximum physical output of the facility. However, since that output cannot be stored (that is, it can only be consumed at the time an activity is undertaken, on site, with the consumer being an integral part of the output process), the actual type and amount of consumption is largely determined by the recurrent inputs; the staffing, management and operation of the facility. Because of this an expansion in output does not require a proportional expansion in inputs until, at the margin, the capacity of the original inputs is reached. This was found to be at 56 000 uses per annum for the small sports halls developed under the Sports Council's SASH scheme that were used as the basis for developing this technique. In order to expand the consumption beyond 56 000 annual uses more recurrent inputs, such as staff, are required. Eventually the total capacity of the facility will be reached, where the only solution to a continued expansion of output is a new injection of capital-inputs to increase the size of the facility.

This analysis of the input-output relationship has largely equated consumption with output. Dependent upon the aims of the provider, however, consumption could be a poor indicator of output, especially where those aims include provision for minority groups, or the provision and promotion of different activities. Yet, as has been found with the other methods of evaluation, there is no recognised way of evaluating these non-financial objectives and the consequent output of the facility.

During the fieldwork it was found that a relationship exists between the objectives as intentions and the nature and magnitude of the inputs used. In particular, the intention to concentrate on certain target groups or activities can fashion the type and size of the facility as well as the style of staffing and management. Once the facility has been developed to cater for certain intended use and users, no expansion of inputs would be necessary until the target output has been reached. This means that the cost of the facility up to the target output is a function of intended, rather than actual output.

The results from the study of the sports halls indicate that differences in intended final output, ranging from a strong welfare orientation to one based on the competitive market, can lead to differences of up to 20 per cent in input costs for otherwise similar facilities. Furthermore, over the nominal capacity of 56 000 uses per annum, marginal additions to consumption require marginal increases in inputs and, therefore, cost (see Figure 4.6). This means that the manager of an individual facility has the ability to make some form of objective evaluation of the performance of that facility, in both absolute and comparative terms. Thus, the basic annual cost for a facility with 56 000 uses per annum should be in the region of £165000, with a marginal addition in cost of £2.74 per use up to a maximum of 68000 uses per year. If the actual costs for a facility are higher than these, the first step is to consider the degree to which the objectives for its operation reflect a social or welfare orientation. If they do, the annual staff and operating cost categories can be deflated by up to 38 per cent, depending upon the degree of welfare orientation, in order to see whether they fall into line with the 'standard' cost. If they do not, or if there is no pronounced welfare

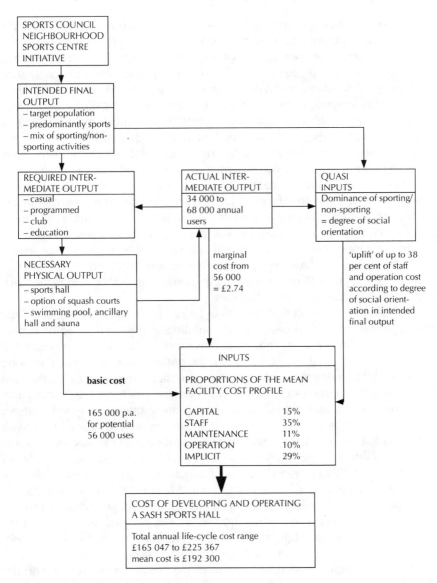

Figure 4.6   Determination of the cost of developing and operating a SASH sports hall (Figures as at 1985)

orientation to the objectives anyway, the manager should examine the operation of the facility to ensure that it is being run as efficiently as possible.

These findings, established for one type of leisure facility, illustrate the type of input-output relationship to be found in leisure development and the way in which it can be used to evaluate performance. The findings are not, however, capable of explaining this relationship yet, due to the continuing inability to measure non-financial outputs in finite terms, according to their own value systems. However, indications are that the relationship between the inputs and outputs of non-profit recreation provision may be every bit as direct as that for the profit-orientated sectors. It shows, furthermore, that the prerequisite for this relationship to exist is not an ability to compare inputs and outputs in a similar value system, as is the case with cost benefit analysis, but rather, to be able to measure each according to its own value system, so that changes in the value of one can be compared to changes in the value of the other. Until this ability exists, however, output intention can be used to formulate a comparative assessment of output values, from least-to most-welfare orientated. This ordinal system can then be superimposed on the input value system with, according to the work on sports halls, a multiplier of up to 38 per cent of staff and operation costs between the least- and most-welfare orientated facilities.

In putting this solution forward, however, it must immediately be recognised that it is, at best, a partial solution, since a unique value system for output does not exist. The comparative ordering of output intentions that has been developed as a surrogate for the output value system is, therefore, an attempt to refine the existing monetary input value system to compensate for this lack of output value system, but certainly not to replace it, or to give the impression that it represents an appropriate output value system in itself. Indeed, extreme caution must be exercised in its use, particularly with respect to decisions concerning the appropriate output classification for any given facility and the implicit assumption that increased costs are always associated with an increased welfare orientation. In the former case the link between output option and input cost could become self-fulfilling, either by the operators of high-cost facilities claiming highly social outputs, or by the selection of social objectives followed by the demand for increased re-sources simply to attain those objectives. Equally, in the latter case, increased commercial activity could have extra costs associated with it. In all cases, therefore, the development of the measurement system and the selection of the appropriate output option and associated multiplier must be kept entirely separate from the policy formulation and operation of that facility.

In concluding this subsection, it has been shown that whilst a range of techniques has been derived in an attempt to place a value on non-priced recreation facilities, objections can be raised against any of them. However, as Bovaird (1991) argues, the fact that objections are raised should not lead to the techniques being discarded, particularly when many of the objections can be circumvented or, at the very least, acknowledged and, to a degree, accounted for.

4.4.5 Funding

Whatever the project and regardless of the eventual user, the actual development of the site or facility will have to be funded, in part at least, in advance of completion and long before any substantial return can be made. It is, therefore, important to know who is responsible for arranging the funding and what risks they are incurring. Generally in the speculative world of commercial development, the funding will be the sole responsibility of the developer:

> In simple terms, a commercial/industrial property developer will purchase land and construct buildings on short-term finance borrowed from a bank and then let the accommodation to one or more tenant occupiers on the basis of a twenty or twenty-five year, full repairing and insuring lease with five-year upward only rent reviews. This stream of net income is then sold as an investment to one of the long-term property investors (Cadman 1984, 76–77).

In the case of the speculative developer the risk in obtaining funding lies in the accuracy of building cost predictions and the appraisal of the potential client market. Since there is no client prior to the development commencing, there can be no client's brief to follow and, hence, no chance of a disappointed or defaulting client.

Since speculative development is relatively rare in recreation projects, most development is undertaken by a few specialist developers on behalf of local authorities, commercial leisure operators or the voluntary sector. In these cases, where the client commissions the development, the risk is associated with selecting an adequate developer as well as undertaking an appraisal of the potential use of the new facility.

Because of the non-speculative nature of the market and the range of clients seeking funds, there is, at best, a very poorly developed system of funding recreation developments, notwithstanding that some similarities do exist with the funding of other types of property development. In the past, recreation development has been partially dependent upon government grant aid, certainly as an indication of acceptability. For countryside projects this could amount to grants covering half the development and operating costs, payable from the Countryside Commission. Whilst not being so generous, substantial sums of capital and operating expenditure were also available for appropriate projects from the Sports Council and the national tourist boards. More recently these government derived sources have diminished. They have been replaced, in part, by surrogate aid in the form of planning gain, where property companies are prepared to fund recreation development in order to secure lucrative planning consents.

Beyond actual leisure companies, such as Rank, Ladbrokes and Mecca, for example, other sources of funds for recreation development include the major breweries, either directly or through wholly-owned subsidiary development companies. In either case the funding can vary from direct involvement to low interest loans. In the latter case the loan is usually for developing, renovating or acquiring licensed premises, with the loan tied to the eventual sale of that brewery's products.

For the independent developer who does not wish to be tied to, or associated with, breweries or non-leisure planning consents, finance can be arranged through the banks. The clearing banks will generally lend short term development finance for most types of property. This is usually conditional upon there being sufficient security, in the form of freehold or long leasehold interests. Merchant banks will also arrange funding for leisure development, but usually only for an enhanced rate of interest or an equity share.

However, whilst the banks may have softened their attitude to leisure development, the equity market has remained largely sceptical. The exception to this has been in cases where there is asset backing. A prime example of this has been the predominance of leisure-related development proposals under the Business Expansion Scheme, where companies intending to buy and renovate pubs, restaurants and hotels have proved most popular. Apart from these opportunities, realistically limited to fairly small schemes, the problem of funding large-scale leisure development has remained. For this reason a new venture capital company, Electra Leisure, was started early in 1990, with £40m equity, to fund leisure development. As a measure of the latent demand for funds, over 70 unsolicited proposals were received in the first weeks of trading, with requests for between £1m and £300m of funding (Terry 1990).

Traditionally, local authorities have been in much the same position as the private sector over funding. Apart from any allocation of funds through their capital budgets, local authorities can raise finance in much the same way as a private developer. In addition to the sources listed above, a local authority can approach the Public Works Loan Board, or apply for Basic Credit Approval from Parliament to allow it to raise external funds. Prior to the Local Government, Planning and Land Act 1980 arrangements such as the sale and leaseback of development sites were popular. Under these schemes the authority would sell some land with planning permission for a leisure facility, together with a commitment to lease the completed development. A commercial company would buy the site, build the facility, claiming all tax allowances available to it, and then let the finished building back to the authority. In essence this produced a model similar to the standard speculative building development, without the speculative risk, but with the builder able to claim allowances not open to the authority, as a non-tax payer, whilst the authority could avoid showing any capital acquisitions in its accounts. After the enactment of the statute, whereby the capital element of the leases had to count against the authority's capital allocation, and a gradual reform of capital allowances, most of the advantages were lost (Cadman and Austin-Crowe 1983 and Barclay 1990).

Apart from these methods of funding, local authorities can establish development companies which are then free to raise equity finance on the stock market. Money can also be raised outside the UK, either from foreign lenders or EC grants and loans. Finally, as a part of government, local authorities have the power to raise funds by the issue of stock, debentures, annuities, bonds and bills. However, with the onset of compulsory competitive tendering and the consequent contracting out of services, many local authorities are seeking companies willing to both develop and subsequently operate facilities for them.

Evidence of the future points to conflicting trends, with high interest rates and the looming threat of an oil-induced world recession creating a tough environment for any investment or development, but where leisure related spending is still expanding, so creating a continuing demand for new and better facilities. Under the recent Conservative governments the emphasis has been very much on market-based solutions to recreation provision, in both the public and private sectors of the economy. On the one hand this has led to the development of many exciting facilities, often as a result of public/private joint schemes, with much new funding brought into leisure provision. On the other hand, it certainly appears to be the case that there is less funding available for the relatively simple recreation facilities traditionally provided by the public sector.

4.5 Construction

Whilst 'development' and 'construction' are sometimes thought of as synonymous, construction really represents the final phase of the development process. Also termed production and implementation, the construction phase represents the fruition of the original development intention:

> Towards the end of the development process, after the assessment of market demand and the arrangement of finance, comes the actual design and construction of the project (Fothergill *et al.* 1987, 51).

However, apart from representing the final stage of the development process, construction can also represent the first major commitment by the developer. This is usually the purchase of the land, either in terms of a freehold or a long leasehold interest. It is at this point that the location decision is finally made and funds secured to underwrite the purchase and construction of the facility. Once this stage is reached the developer is unlikely to be able to pull out without risking substantial financial loss.

Once the right to the land has been secured the next step is to place the building contract. This represents the ultimate commitment to the scheme (Cadman and Austin-Crowe 1983). There are three principal methods of achieving this: a standard form of building contract, such as that prepared by the Joint Contracts Tribunal and known as a JCT contract; a management contract; or an especially prepared design and build contract. The first of these is widely used, with the developer being responsible for the preparation of a detailed design and associated Bill of Quantities from which builders may calculate their tender price. In the case of the management contract, the project is managed by a contractor who does not undertake any work directly. Finally, the design and build contract is based upon the developer preparing a detailed performance specification describing the various requirements that the facility should meet, and the builder preparing a detailed design which is capable of achieving that specification (Turner 1991 provides a detailed summary).

Whilst the JCT standard form of contract is widely used, it has attracted its share of criticism. This is mostly based upon the need to produce good drawings and bills of quantities so that there are no unforeseen problems and no need to renegotiate on the basis of a modified drawing. The ability to renegotiate means, however, that flexibility does exist to make alterations in the light of the development as it progresses. It is usual to put JCT contracts out to selective competitive tender, although there is no reason why this need be so, especially if the developer knows of a good reliable builder. This latter method does, however, eliminate any element of competition. Increasingly, regardless of whether the contract was tendered or not, developers are requiring builders to take out a Performance Bond as an insurance against non-completion of the work. This is because of the all-too-frequent occurrences of builders going bankrupt. Although the parties are free to agree any methods and staging of payments, the most common is on a monthly basis according to the value of the work done, and materials on site, less a retention, with the value being assessed by an independent quantity surveyor.

The design and build contract is most appropriately used for simple straightforward buildings where a standard design can be applied to more than one location. This type of contract is often used for farm buildings which, whilst not always straightforward, do tend to be of standardised design and construction. Similarly, this concept was applied to the Sports Council's Standardised Approach to Sports Halls (SASH) initiative (see earlier in this chapter) in an attempt to produce simple 'off-the-peg' sports facilities that could be easily and quickly developed for a known price. Indeed, the issue of economy features strongly in the advantages of design and build contracts. This is because the builder should be able to work more efficiently and speedily and, therefore, more economically (Turner 1991), whilst there may well be some economies in buying materials as well (Stephens 1983). The price for these contracts is usually agreed before building starts and, especially for smaller jobs, is not paid until completion. Whilst this delay in paying is good for the cashflow of the developer, it does tend to mean that the developer has little influence or control of the development once it is agreed and underway.

Regardless of the type of contract that is agreed, the developer still has to maintain control of the development to ensure that it proceeds satisfactorily (Freeman 1989). On larger contracts it may be advisable to retain a project manager to oversee the work on behalf of the developer. In all cases, however, it will be necessary to know how long each stage of the process should take so that any delays can be identified early and rectified as soon as possible.

The nature of leisure development is such that there are relatively few large organisations contracting out much work, but many organisations, both public and private, that contract out individual projects that can be of considerable value. This presents the difficulty that few of these companies or authorities have much experience in organising building work. As Fothergill *et al.* state, with respect to the analogous world of industrial property development:

> Their experience lies in the design and production of the goods they sell, not in property development, and only the large firms have the necessary in-house skills. Few have the know-how to demand 'fast' times, to make informed choices about the organization of their project and to commission it in a way that assures them of all their requirements. Moreover, since in all but the largest firms major building projects are undertaken infrequently, there is little or no possibility of a manufacturer accumulating experience or learning from past mistakes. As things are, inexperienced customers are at a disadvantage because they have no certainty of getting maximum service from the building industry (1987, 53).

4.6 Monitoring and evaluation

4.6.1 Evaluation of the development process

Many authorities believe that the development process is at an end on the completion and disposal of the building or facility (Barrett *et al.* 1978). In these cases an evaluation of the development is largely concerned with an appraisal of the developer's profit. This can be in terms of a capital sum on disposal of the freehold or long leasehold, a rental sum achieved on the letting of the building or facility or, where the building or facility is built for a known client, the surplus left from the agreed contract price once all costs have been met.

In the past the property development industry has been particularly poor in utilising formal methods of appraisal and evaluation, even where such methods are in common use in other industries. In part, this has been a reflection of the nature of property where, once land has been acquired and a building started, the location and physical nature of the development are fixed in the prevailing social and economic framework and are, consequently, largely beyond the control of the developer (Byrne and Cadman 1984). A further reason lies in the finality of the development process, where even if evaluation highlights poor initial appraisal nothing can be done to rectify it:

> As the process takes place, the developer's knowledge of the likely outcome increases but, at the same time, the room for manoeuvre decreases. Thus, while at the start of the process developers have maximum uncertainty and manoeuvrability, at the end they know all but can do nothing to change their product, which has been manufactured on an essentially once and for all basis. (Byrne and Cadman 1984,5).

The basis of the evaluation of a development project is, by its nature, financial. The main constituents of the appraisal are the cost of the construction and site works, the cost of acquiring the land, the professional fees on purchase, design, letting or sale and the cost of borrowing development finance. The principal ratios used to assess the performance of the development project are the developer's yield and profit. The calculation of these is shown in Figures 4.7 and 4.8 using, respectively, hypothetical developments of a golf course and hotel and a combined retail and leisure complex, both for eventual sale as investments.

In both examples it is assumed that the developer has to acquire the land and undertake all development. In practice this would often be undertaken in partnership in order to spread the cost and, therefore, the risk of the development. Equally, on greenfield sites such as those usually used for golf course construction, the partner will often be the landowner, meaning that finance does not have to be found to acquire the land prior to development. Given that the land has been acquired, however, interest will be payable. Even assuming a highly competitive rate of 15 per cent, rather than the 19 to 20 per cent that might be charged by a merchant bank, the finance charges are likely to add over 20 per cent to the cost of the land during the time taken to complete either of the developments. The same will be true for the cost of construction, although not all construction costs need be incurred before development commences. Finally, a letting cost is included. This could equally be a selling cost, or it could be dispensed with altogether if the buildings were commissioned by their intended occupants.

In the first example (Figure 4.7), therefore, the total cost of development is estimated to be £5 926 250. Using comparable evidence the developer is able to estimate that the likely annual rental is in the order of £500 000. However, with so little available evidence in the leisure sector there could be substantial error in this figure. Using these two figures, the Developer's Yield is seen to be 8.44 per cent. Again, little comparable evidence exists, but a Yield of closer to 10 per cent might be expected in other development sectors.

In considering the capital value of the development it is usual to use the investment method of valuation. However, again the level of comparable evidence is slight, thus necessitating a reliance on figures for non-leisure development. Expected yields for prime commercial developments would be in the region of 7 per cent, with those for lesser developments being closer to 9 per cent. Whilst it would be unwise to equate leisure with commercial property, a golf and hotel complex might currently be considered a prime leisure development, especially in a good location. Even so, it is unlikely that it would be considered to be wholly comparable with the prime commercial development, thus suggesting an investor's yield of 8 per cent. Multiplying the yield by the rental value produces a capital value of £6.25m, which is just in excess of the total development cost. This differential produces a small Developer's Profit of just 5.5 per cent, as

LAND COST		
Freehold with vacant possession		
150 acres (60 ha) @ £10 000	1 500 000	
cost of acquisition – agents fees		
stamp duty, legal fees @ 4%	60 000	
	£1 560 000	
Interest on land held from date of		
purchase to disposal, say 18 mths		
@ 15%, say	350 000	
total land cost		£1 910 000
CONSTRUCTION COST		
75-bed hotel @ £25,000 per room	1 875 000	
18-hole golf course @ £50,000		
per hole	900 000	
	£2 775 000	
professional fees @ 15%		
inc VAT	416 250	
total construction cost		£3 941 250
LETTING COST		
Agent's fees @ 15% rental income	75 000	
		£75 000
total development cost		£5 926 250
EVALUATION		
Total Development Cost		£5 926 250
Rental Income, assume		£500 000 pa
Developer's Yield		8.44%

$$\frac{\text{rental income}}{\text{total development cost}} \times 100$$

Capital Value		
on basis of investor's		
yield @ 8% x rental value		£6 250 000
Developer's profit,		
capital value less development		
cost, expressed as a percentage		
of the total development cost		5.5%

Figure 4.7 The speculative development of a golf course and hotel

against Developer's Profits for commercial property that may, in a buoyant economy, be in excess of 20 per cent.

As a comparison, the mixed retail and leisure complex outlined in Figure 4.8 represents an increasingly popular form of commercial development, where the leisure element ensures not only a fuller use of the site, but also entices a wider range of people to the complex.

LAND COST
Freehold with vacant possession
4 ha @ £750 000 .... 3 000 000
Cost of acquisition – agents' fees
Stamp duty, legal fees @ 4% 120 000
                                              3 120 000

Interest on land held from date of
purchase to disposal, say 18 months
@ 15%, say 700 000

             total land cost   £3 820 000

CONSTRUCTION COST
retail space 6 500 m^2 @ £500 .... 3 250 000
cinema 3 000 m^2 @ £700 .... 2 100 000
ice rink 800 m^2 @ £750 .... 600 000
restaurant 400 m^2 @ £600 .... 240 000
                                              6 190 000

professional fees @ 15% inc VAT 928 500

             total construction cost   £7 118 500

LETTING COST
Agents' fees @ 15% rental income 153 375

             total development cost   £11 031 125

EVALUATION
Total development cost   £11 061 500

rental income

assume £110 per m^2 retail 812 500
assume £50 per m^2 leisure 210 000

                                      1 022 500 pa

Developer's Yield
$$\frac{\text{rental income}}{\text{total development cost}} \times 100$$
                                                    9.3%

Capital Value
on basis of investor's yield
@ 8% x rental value   £12 781 250

Developer's Profit
Capital value less development cost,
expressed as a percentage of the
total development cost   15.9%

Sources: Davis Langdon and Everest (1992); Smith (1988)

Figure 4.8 Development of a retail and leisure complex

In contrast to the golf development, far less land will be required, although an accessible urban fringe location ensures that the total cost of land will be twice that of the golf development, at nearly £4m. Construction costs will be highly dependent upon the degree of fitting out that is undertaken prior to sale. It has been assumed in Figure 4.8 that all units will be ready for occupation, at a total cost of over £7m.

Rental figures are extremely hard to estimate for the initial opening periods of retail complexes since large incentives may be necessary to attract the desired mix of shops. Indeed, since the commercial success of the complex is likely to pivot on the ability of the letting agent to attract a large food store, it may be necessary to offer rent free periods of up to several years. Equally, what is evident is that the leisure elements of the complex are unlikely to be able to compete in rental terms with the retail units. For this reason the figures of £125 and £50 per square metre have been selected for retail and leisure respectively as being indicative of their likely magnitudes.

Using an investor's yield of 8 per cent in recognition of the poor economic climate and the retail/leisure mix in this development, a Developer's Yield of 9.3 per cent and a Developer's Profit of 15.9 per cent are achieved; both being comfortably in excess of the corresponding figures for the golf/hotel development. Although better than the figures for the golf/hotel development, they remain low for such a costly and speculative venture, primarily because of the high cost of land and construction compared to the relatively low rents chargeable at the start of the scheme. Equally, it is tempting to conclude that the leisure elements are compromising the commercial viability of the scheme since they cost more to construct and are achieving a much lower rent. Whilst this is partially valid, however, there is no indication of how the absence of the leisure elements would have affected the retail rents.

With such a low Developer's Profit it is highly unlikely that the first scheme, involving the golf course and hotel, would be undertaken speculatively; indeed, it underlines the fact that a great many leisure facilities apparently have capital values that are lower than the total cost of development. It further suggests that some of the claims as to the profitability of this type of leisure development are grossly exaggerated, or based on incomplete data. In particular, ignoring the cost of the land, whether in terms of actual money or opportunity foregone, can have a profound effect on the financial success of the development. If, in the example in Figure 4.7, the land was already owned by the developer and not charged against the development, the Developer's Yield would be nearly thirteen per cent whilst the Developer's Profit would be a massive fifty-nine per cent. Equally, charging the land at its value for the next best use could also be misleading, particularly if that use is agriculture or forestry.

In contrast, the mixed retail/leisure development could be contemplated on a speculative basis, as long as there was a reasonable chance of securing some prestigious retail tenants for the commencement of the scheme. In terms of the leisure input, however, it is patently obvious that it could not be contemplated separately from the retail element and that it may still be jeopardising the success of the entire complex.

In conclusion, evaluation of the development itself tends to be limited to the simple financial concepts of Developer's Yield and Profit. These are generally estimated as part of the tendering and feasibility phase of the development process, with commencement dependent upon sufficient yields and profits being achievable. At the end of the process, therefore, evaluation is really a case of relating actual performance to that planned. Furthermore, since many of the cost items were of known magnitude at the commencement of the development, the process of evaluation is really concerned with cost control at the construction stage and the ability to let or sell at the estimated price.

4.6.2 Evaluation of the operation of a leisure facility

The final stage in the evaluation of a recreation facility, although not commonly thought to be part of the development process, is the measurement of operational performance. The purpose of performance measurement is to assess the operational management of the facility. In the commercial sector this is usually in the form of financial evaluation, but should also encompass data on user numbers, perceptions and other non-financial indicators. In his work on financial management in leisure provision, Stabler (1990) illustrates the use of accounting ratios to measure performance. The most common type of ratio is based on profitability, where the central relationship is between capital invested and profits generated. Other ratios may link important factors such as sales, stocks, salaries and value added but, as Stabler points out, the relationship between capital and profits is the most fundamental:

> What might be termed the 'bottom line' ratio is the profit/capital ratio which enables the rate of return on capital, as a percentage, to be established. In economic terms this should at least reflect the opportunity cost of funds invested in the business, perhaps the prevailing market rate of interest being the datum (1990, 112–113).

The second set of performance criteria concern solvency and liquidity. These deal with the structural relationship between assets and liabilities such that solvency represents the ability to meet short term liabilities from income, whilst liquidity deals with the ease with which assets can be converted to cash in order to meet liabilities. Unlike the profitability ratios, these ratios are only of importance to the commercial operator in indicating the financial health of a business, and have little application to public leisure provision.

In contrast to the commercial sector, where the monitoring of performance has always assumed considerable importance, performance assessment in the public sector has really only gained momentum in the last decade, and even then more in terms of discussion than implementation (Bovaird 1991). In particular, the 1980s have witnessed Value for Money approaches to performance assessment (Butt and Palmer 1985) and, more recently, approaches based on the concepts of Quality Assurance and Total Quality Management, as defined in British Standard BS 5750. Value for Money assessment, as laid out in the Financial Management Initiative launched by central government in the mid-1980s, is comprised of four elements: economy in the purchase of inputs; efficiency in minimising the inputs per unit of output; effectiveness in increasing the welfare of the community through the achievement of objectives; and the organisation of management to ensure that economy, efficiency and effectiveness can be maintained in the long term (Bovaird 1991).

In considering each of these, economy indicators would include the purchase price of standard inputs, or a range or 'basket' of standard inputs (Bovaird 1991). These costs could be compared between different parts of an authority, or between neighbouring authorities. Indicators of efficiency include ratios such as output or activity per member of staff, unit costs of output, achievement of agreed standards or the meeting of agreed deadlines (Bovaird 1991). Although these first two indicators are specifically related to the public sector, it would obviously be within the interests of the commercial sector to establish and monitor similar ratios, particularly in situations of limited capacity, where the ability to control costs may be of paramount importance. Equally, these types of measures have become popular with local authorities preparing facility management contracts under the compulsory competitive tendering arrangements imposed by central government.

Performance measurement begins to differ radically between public and private sector when the question of effectiveness is addressed. As outlined above, effectiveness in the private sector is essentially defined in financial terms, using ratios of the type described by Stabler (1990) above. In contrast, effectiveness in the public sector is the combination of the throughput of people affected by the facility or service and the impact of that facility or service on each individual person (Bovaird 1991). Measures of the throughput are primarily concerned with numbers of different users, weighted according to priority groups, together with those indirectly affected by the nature or availability of the facility or service. Indicators of the impact of the facility or service would then include changes in user satisfaction and welfare, measured directly through surveys and indirectly through statistics on the incidence of crime and vandalism in the locality (Bovaird 1991).

Whilst none of the concepts or techniques of Value for Money performance assessment is either particularly new or innovative, their relevance has been heightened by the introduction of compulsory competitive tendering, since local authorities are now extremely keen to make sure that the quality of their services is not allowed to slip once management has been handed over to contractors (Audit Commission 1990). In achieving this it will be necessary for authorities to develop clear standards related to the required quality of service; indications of what constitutes acceptable quality of service; a planning and design process that will ensure quality standards are maintained; and an agreed procedure for monitoring whether these specifications are present in the actual delivery of the service (Bovaird 1991). Whilst accepting the benefits of this approach, Bovaird does issue a warning about what is actually being measured:

> These steps offer important ways of ensuring that leisure departments adopt "quality management systems"; practical approaches can be and are being devised to achieve this in many local authorities. While they help in assessing quality, we should be clear that they do not involve measuring quality. Those even more grandiose schemes which purport to measure quality in quantitative terms are chasing fool's gold. (1991, 16).

Equally, it is important to recognise that quality management systems tend, by their very nature, to relate to facilities and services, rather than to the effect of those facilities and services on their users. In effect, therefore, such an orientation can do as much to ensure that the effectiveness of the facility or service is not improved, as it can to provide the catalyst for a radical shift in management.

In concluding this section it is important to reflect upon the evident lack of importance attached to the evaluation of developments, even where the eventual occupant commissioned the development. In part this is due to the short-term speculative nature of the development process for most types of property, where evaluation is solely concerned with the developer's profit. However, in leisure provision, where the majority of facilities are commissioned, evaluation is still at a primitive stage, even in the public sector where the generation of welfare benefits is paramount. Thus, on the one hand is the private commercial organisation, where finance is available as long as it can be shown that projects can repay the borrowings in a relatively short period of time. On the other hand is the public, budget-based organisation, where analysis is undertaken to determine the necessary size of the initial budget, but once allocated, the budget must last a specific length of time. Initiative exists in determining how to spend the budget, but the decision as to which projects are financed is dependent upon the size of the budget and the initial forecasts of financial needs.

Organisations representing either of these models require detailed initial financial estimates. Those representative of the first method need to deflate costs and inflate income as much as possible in order to attract the finance they require at the most preferential rate possible. Conversely, the second method requires inflated costs (within the anticipated limits of the budget) in order to secure the largest budget possible for the project. The performance of the first type of organisation will, therefore, be based largely upon its ability to service the borrowed finance and return an acceptable level of profit to its owners. In contrast, the performance of the latter type of organisation, characterised by a local authority recreation department, will be based upon the ability to maintain or increase the size of budget (Drucker

1977). Therefore results, in terms of meeting welfare objectives, can become secondary. Whereas the commercial enterprise is reliant upon end-of-year financial returns, public organisations rely on the formulation of good intentions and substantial budget requirements to attract a sizeable budget, a large enough programme to ensure its use and as little subsequent performance evaluation as possible. When performance is largely budget-based, the organisation concerned is encouraged to spend as much money as it is allocated simply to justify the continuation of its budget. Even the recent shift in governmental attitudes to performance assessment does not appear to have made any appreciable difference:

Value for money studies have thrown up quite a number of one-off reports of economy, efficiency and effectiveness in leisure departments. However, they have apparently not led to marked changes in the way in which departments think. The greatest culture change in leisure departments currently appears to be associated with the need to respond to CCT (*Compulsory Competitive Tendering*); yet this would not have posed major problems if there had already been a tradition of searching and analytical performance measurement of services, feeding back into a service appraisal system which sorted out the best services in advance of delivery (Bovaird 1991, 26).

5 State Regulation of Development: the Statutory Planning System

5.1 Introduction

The most significant part of the development process not analysed in Chapter 4 is the influence of state regulation and control of development. For the majority of the Twentieth century the most fundamental aspect of state intervention in the development process has been guidance, through the preparation of development plans, and regulation, through development control. In addition, the state has also developed and enforced various regulations about the construction of buildings and the use of the environment.

In technical terms, consideration of planning issues is part of the evaluation process in determining whether a proposed scheme is likely to gain the necessary development consents. It is, therefore, critical for a prospective developer to be aware of the operation of the planning system in order to expedite consent, or avoid the preparation of costly schemes that are never likely to be given the go-ahead by the planning authority. In more philosophical terms, the underlying role and function of the town and country planning system is to intervene in the free market mechanism in order to control private entrepreneurialism and safeguard the public against inappropriate uses of land. As such, the system is bound to provoke controversy and dissent since its existence and use has implications for the otherwise largely unencumbered use of land and property, as described in Chapter 3. Thus, the town and country planning system is much more than a mere tool for circumscribing individual freedom. Rather, it is a means of mediating between public and private interests, whether to protect the public from the excesses of the market mechanism, or to protect private individuals from ill-considered public development, although in this role it has been somewhat less effective.

The theoretical justification for this intervention in the free market has been based on three factors: aesthetics; functionalism; and externalities. Although seemingly the least important of the three, as well as the least well defined, a range of aesthetic considerations has been the consistent driving force behind town and country planning throughout the Twentieth century, particularly in the containment of urban development. Furthermore the recent growth in leisure development has returned aesthetic issues to the forefront, in debates over the future of the countryside and the place of 'non-rural' development within that countryside.

Using planning to make things work better has long been a justification for the statutory planning system, particularly in the functional relationship between housing development and the provision of schools, hospitals, shops and the workplace. The functional justification of planning was probably at its height during the inter-war years of mass house building and before many households could afford a private car. However, it was also one of the central reasons why the Regional Councils for Sport and Recreation were charged with the duty of having recreation provision considered during the development of the new Structure Plans in the early 1970s, so that access to such facilities would be readily available to all.

Finally and, arguably, most importantly, are the externality justifications for statutory planning. In common with the externality arguments for the public provision of recreation facilities (see Chapter 7), the economic rationale for statutory planning centres on the need to provide certain public goods (such as roads, street lighting, the police and emergency services and, more contentiously after the last decade of government, education and health care), to circumscribe the effect of certain negative externalities by limiting the location of dangerous or noxious industries, and to protect or enhance the provision of certain merit goods such as public open space and the density and layout of housing schemes.

Implicit in all these justifications of statutory planning is that a socially inefficient allocation of land uses would occur were it not for the moderating influences of the system and its operators:

> The existence of town and country planning seems to be predicated on the assumption that there is some pattern of land use which is socially desirable, but which is different from the pattern which either the market, or the ... other types of government intervention ... would produce. (Reade 1987, 3).

However, it is difficult to know how far the present land use pattern found in Britain is the product of the town and country planning system, or has been shaped by other forces. Quite apart from the large stock of buildings and uses inherited from former centuries, the vast majority of the country has never been subject to any appreciable development pressure, whilst repeated attempts to relocate people and their work away from the heavily crowded south-east of England has, for the most part, failed. The situation as far as planning for recreation is concerned is even more difficult to determine, given the wide range of public, private and voluntary suppliers. What is certain, however, is that recreation has never featured as a central part of the town and country planning system, suggesting that the current pattern of provision may have very little to do with the statutory system at all. The remainder of the chapter will start with an overview of the present system of development planning and control, followed by a discussion of the evolution of the system and the place that recreation planning played in that evolution. Chapters 6 and 7 will then build on this overview of statutory planning by relating it to recreation development in, respectively, rural and urban areas.

5.2 Development plans and development control

The basis of the current planning system are the twin concepts of development plans and development control first introduced in the Town and Country Planning Act 1947. Now modified by the Town and Country Planning Acts of 1971 and 1990 and the Planning and Compensation Act 1991, the development plans consist of Structure Plans and different types of Local Plan, whilst the basis of control consists of the expropriation of an owner's right to develop land without obtaining prior approval from the planning authority. Since the legal definition of development includes any physical alterations or changes in the use of land (see Chapter 3), this expropriation really covers all uses and operations apart from the current one.

The basis for determining whether or not a particular application should be approved are the Structure and Local Plans, together with any other material considerations, allied to the judgement of the planning officers and elected council members. Since the planning system operates on the basis of administrative discretion rather than rule of law, statutory plans can provide no more than guidance to a prospective developer. Thus, whilst there may be statements of planning objectives backed up by Development Plans, there is also a large amount of discretion to consider each case on its merits. Although providing a flexibility not present in systems with less discretion, the British system is heavily reliant on professional planners to interpret planning policy, thus leaving it potentially open to selectivity and inconsistency.

Structure Plans are drawn up by the County Planning Authority and comprise a written statement of policy and associated diagrams. They are, in effect, plans without maps (Cullingworth 1988). Rather than the concentration on land use in the earlier type of development plans, the Structure Plan is concerned with major economic and social forces and how the County proposes to deal with them in land use terms. Within this framework the County may outline Action Areas which indicate where major changes are likely to occur and which should receive special attention in the preparation of Local Plans. The Structure Plan must be approved by the Secretary of State for the Environment before it can become a statutory document.

Local Plans do not need Ministerial consent since they are supposed to keep within the framework laid down by the Structure Plan. The first type of Local Plan is the General Local Plan, covering a relatively wide area such as a small town. A General Local Plan will cover all aspects of planning policy for that area, as well as the relationships between different areas of policy.

Subject Plans, on the other hand, deal with specific areas of policy, such as housing, industry or, very occasionally, recreation and leisure (Hill and Healey 1985 give examples).

Although planning decisions are ultimately made by elected representatives, developers usually start their application process by holding discussions with planning officers. The outcome of these discussions is often a modification of the intended submission in order to take account of the officer's views, both on the intended development and the likely views of the councillors themselves. Once submitted, the applications are considered by the councillors under the guidance of the planning officers. Whilst the councillors have the power to make a decision, the Secretary of State for the Environment can 'call in' contentious proposals and make the decision on behalf of the planning authority.

Three decisions can be made: approval; approval subject to conditions; and refusal. The applicant has a right of appeal to the Secretary of State in the latter two cases, with the appeal being dealt with by an inspector appointed by the Secretary of State. Appeals can either be by written representation or, in the more complex cases, by public local inquiry. A public local inquiry can also be used when the Secretary of State for the Environment 'calls in' an application which is either highly complex or at odds with the development plan policy (see Chapter 9 for an example of a proposed recreation development going to public inquiry).

Apart from the power to grant or refuse planning permission, the local planning authority also has the right to serve a Stop Notice to halt any unlawful development until there has been time to consider it, and an Enforcement Notice which compels the developer to demolish or remove whatever has been developed.

Whilst it is the general case that every development or change of land use requires permission, there are two types of exception, covered by the Use Classes Order and the General Development Order. The former of these prescribes classes of use within which a change can take place without it constituting development. However, this refers solely to changes of use and not to any building works. The latter prescribes classes of development for which permission is not required. The largest class of development to which the General Development Order applies is agricultural where, with certain restrictions, farmers do not need permission to construct new buildings or carry out development works. On a smaller scale, all householders have the right to increase the floor area of their house by up to ten per cent per year without the need for planning permission. In both cases, however, there will be a need to obtain Building Regulations consent and satisfy the local authority buildings officer that this has been adhered to in the construction.

Of most relevance to recreation development are the short-term permissions contained in the General Development Order. These allow certain uses, not including physical development, to be carried on for a maximum of either fourteen or twenty eight days per year without the need to apply for permission. The most common of these are fairs, fetes and travelling circuses which are only likely to stay on the same site for a few days at a time. More permanent, but not requiring permission for less than fourteen days use per year are clay pigeon shooting sites, motor cycle scramble tracks and other 'noisy' sports.

The use of sites for caravans and camping can also fall within the ambit of the General Development Order. Whilst major sites that will operate continuously throughout the year will need planning permission, as well as an operator's licence from the local authority, sites run under an agreement with one of the national caravan or camping clubs may not. To qualify for the latter case the site must be restricted to club members only, be limited to five caravans or tents per night, with no individual caravan or tent remaining on site for more than twenty eight days per year.

Whilst most of these excepted uses are minor and inconsequential, recent proposals by the Conservatives to extend them met with fierce opposition from many quarters. The basis of these counter-arguments was that although fourteen days per year was not many, it still enabled noisy sports to operate on the same site once per week for most of the summer; any extension would simply exacerbate the problem. In the face of these objections the government was forced to back down, retain the existing controls and even propose to expand them (Byrne and Ravenscroft 1989).

In conclusion, therefore, the planning system as it has developed over the last century is essentially a conservative set of measures designed to protect the interests of existing landowners from unwanted outside pressure. This is achieved through a mixture of administrative discretion related to develop-ment plans and locally elected planning authorities advised by professional planners. Whilst this system has proved most successful in containing and concentrating development, it has proved to be somewhat inflexible in meeting the changing needs of society and adapting to new types of development as they are proposed. This situation is well illustrated in the following chapters with respect to urban and rural recreation where, for the most part, new developments have only been allowed or encouraged when they meet some particular need of government or are of no consequence to existing landowners and local inhabitants.

5.3 Evolution of the town and country planning system

It is generally agreed that modern town and country planning arose as a result of the rapid growth of Victorian cities. Perhaps the greatest concern was over the link between public and personal health, with Public Health Acts being passed in 1848 and 1875. The particular health concerns were over the ability to provide clean water, to clear refuse and sewage and to treat epidemics (Hall 1975). Similar concerns were expressed for the quality of housing, whilst Smith (1974) makes the point that the rate of growth tended to deprive people of any open space or greenery. However, state intervention in the process of expansion remained limited to the health related issues, with little concern, initially at least, for the wider issues of housing, public open space or the relationship between people's living and working environments.

As a consequence, much of the early work on the planning of urban environments was left to many individuals who developed their own interests and responses to the evident problems. One such person who was to have a profound effect on the later formulation of town planning was Octavia Hill. Her interests in the relationship between housing, education, open space and, consequently, preservation of the environment, started the link between recreation and health that has continued to underpin much public provision for recreation over the last century. Octavia Hill went on to found the National Trust and continue her campaign to ensure adequate recreation opportunities for the urban poor. In a similar vein, Ebeneezer Howard and others were developing their garden city and garden suburb concepts as a means of providing a well-planned environment for both industry and the new industrial workforce.

Further north there were a number of new towns being constructed to house industrial workers. These towns included New Lanark, Saltaire, Port Sunlight, New Earswick and Bourneville. In all cases, as Smith suggests, the towns were built to ensure that workers enjoyed a better and more healthy life than had previously been available in the old cities:

> All the principal schemes which were implemented between 1853 and 1901 were dependent on enlightened entrepreneurs who wished to house their workers in convenient and pleasant surroundings, and all were characterised by relatively spacious, semiformal layouts with provision for greenery ... (1974, 23).

However, the effect of the new development, as well as the continuing growth of existing cities, meant that those who happened to own land in these areas were able to sell it at a profit to the developers. Indeed, as the cities grew, so did the value of the land in and around them. By 1900 this had led to attacks on the 'profiteering' landowners, with many people claiming that these landowners had no right to the windfall gains, or 'unearned increment' that was due to an increased societal demand for land (Reade 1987).

Even more pressing, however, was the need to consider whole areas of development, rather than simply individual properties. In addressing this issue the Housing and Town Planning Act 1909 was passed. In this first formal attempt at town planning, the Act provided for local authorities to produce schemes to control the development of new housing areas in order to secure proper sanitary conditions, amenity and convenience (Allison 1975). Once again, therefore, it can be seen that amenity and recreation were considered to be important elements in the first formal attempts at implementing a statutory planning system. Although primarily passed to address the national problem of improving the lot of the urban proletariat

(Allison 1975), as well as curbing urban sprawl (Reade 1987), the Act addressed, for the first time, the issue of who should own the development value of land. Until the enactment of the 1909 Act the presumption had been that landowners were free to use their land as they wished and to profit accordingly. However, the 1909 Act introduced a betterment tax on the increase in the value of land attributable to a development scheme. In effect this meant that in cases where the increase in the value of the land was attributable to societal demands, the state rather than the individual should benefit. Rather than follow this principle to its logical conclusion by refusing to compensate those whose land value was restricted by the planning schemes, the Act did the opposite, so leaving the local authority liable to pay compensation to those affected by its plans. Since, at that time, the schemes were voluntary there was little chance of local authorities having to compensate owners.

By 1915 over 100 plans had been prepared under the provisions of the 1909 Act, covering nearly 170 000 acres. Because of this success the scheme was extended in 1919 by making such plans compulsory for all local authorities with more than 20 000 inhabitants in their area. So far, however, the plans related to private development, with the local authority merely setting the standards. This was changed in 1924 when the Housing Act authorised the building of council houses, but at a maximum density of twelve dwellings per acre, so allowing ample open space and recreation areas, the standards for which were laid down by the National Playing Fields Association in 1925.

Whilst the housing development plans were succeeding in their aim of promoting a better quality of home (Allison 1975), the question of betterment and who should benefit from such schemes was less clear. Although successive inter-war planning acts reaffirmed the principle of a betterment tax, Reade (1987) could find only three instances up to 1939 where a developer had actually been forced to pay the tax. Equally, however, the acceptance of public compensation for planning restrictions ensured that local authorities did not become too involved in the detailed planning of private developments, preferring to concentrate on 'little more than the securing of adequate standards of layout and design ...' (Reade 1987, 42).

Inevitably this residual role led to the whole system of town and country planning being subjected to scrutiny. Whilst the quality of housing had undoubtedly improved since the 1909 Act, the co-ordination between that housing and work places, leisure facilities and transport was inadequate (Allison 1975), whilst there was clearly no national view or plan for future development (Cullingworth 1988). This situation has led Smith to state that:

> ... it had been realised for some years that planning ... as embodied in the 1909 Town Planning Act in terms of residential-estate layout was insufficient both in concept and in practice. Ironically, town planning had not only woefully failed to deal with the partial urban problem that it had concerned itself with, namely the control of new development, but it had not tackled even the most critical aspects

of this – the balance of functions and facilities that was clearly necessary for a 'healthy industrial community'. (1974, 33).

Through the 1930s there emerged two distinct schools of thought on the future role of planning. At a local level there was still a conviction that the merit good aspects of the relationship between improved physical surroundings and the health and welfare of the nation was its primary function. At the strategic level, on the other hand, there was a growing group of economists, led by J.M. Keynes, who considered the externality arguments for planning to be paramount in directing development to the depressed areas of the country. Despite their differences, however, both groups were in favour of prohibiting further development and urban sprawl in the south east; the former group principally on the grounds that it was destroying valuable open space and the latter on the grounds that new development in the north would yield much greater benefits to the nation (Reade 1987).

There followed a number of government reports into the future of the nation, including the distribution of industry (Barlow 1940), the question of compensation and betterment (Uthwatt 1942) and land use in rural areas (Scott 1942). The findings of these reports led to a wide ranging discussion about the future role of the planning system. In particular, much support was gained for the concept of new towns designed to provide a balanced life for inhabitants. However, perhaps the most significant of these for the development of recreation planning was the Scott Report on Land Utilisation in Rural Areas. Indeed, it is difficult to exaggerate the degree of impact that this report has had on the post-war rural scene. With the exception of a dissenting report from one member of the committee (Lowe *et al.* 1986), the recommendations of the Scott Report were that the production of food should be given absolute priority. In order to facilitate this it was proposed that development in the countryside other than for agricultural or forestry purposes should be severely restricted. When enacted in the Town and Country Planning Act 1947, and combined with agricultural price support under the Agriculture Act 1947, it meant that very little of the countryside was available for recreation or other non-farming functions.

Before any of the findings, with the consequent restrictions on development, could be acted upon, however, the question of compensation had to be tackled. Many solutions were discussed, including the nationalisation of development land. Ultimately, however, the existing compromise of betterment being taxed and compensation being paid was felt to be most appropriate. Other proposals included the need for national parks, nature reserves and areas specifically set aside for an ever-more urban public, a public right to wander over all open countryside, the recording of all rights of way on publicly-available maps and a distinction between 'landscape conservation for amenity and habitat conservation for science' (Blunden and Curry 1990, 49); see Chapter 6 for a fuller discussion of these initiatives.

When the Labour Party was returned to power in 1945 with an avowedly socialist manifesto the time was right for some major legislation to reform the balance of power in society. Whilst the central platform of Labour's plans was the creation of the Welfare State, an important measure in achieving a greater degree of social equity was reform and extension of the planning system. This was achieved in the Town and Country Planning Act 1947. In shifting the balance of power very firmly from the landowner to people in general, the Act removed an individual's right to develop and confiscated all development value in the land. However, in following this course of action Labour also helped existing landowners by protecting all rural land from non-agricultural development. Quite apart from preserving the stately pile, this prohibition on new industry in the countryside had the effect of maintaining the monopoly employment position enjoyed by land-owners and farmers, so restricting the increase in farm workers' wages and retaining the existing class divisions (Reade 1987).

The approach to containing and restricting development in most rural areas was reinforced in many parts of the country where the visual quality of the landscape was regarded so highly that it was given the extra protection of being designated as an Area of Great Landscape Value in the new develop-ment plans. Thus the new county policies acted in support of the existing rural interests which were primarily agricultural and opposed to any pro-posals for change. The one exception to the prohibition on development in the countryside was the concept of the new towns. Separated from the existing towns and cities by green belts of agricultural land, the new towns would eventually permit the reduction of the high residential densities in the cities through relocation. This, in turn, would allow the old housing areas to be rebuilt with much more room for public open space and other amenities.

The planning system devised in the 1947 Act was based on three parts: development plans; development control; and a development charge. In planning the future development of an area, the local planning authorities had to draw up plans combining their projections and strategy for the next twenty years, subject to five year reviews. The plan was to be in the form of statements supported by maps indicating the authority's intentions for development and preservation. All plans were to be submitted to central government for approval, meaning that all planning authorities really had to follow national trends and initiatives rather than develop policies specific to their perception of their local needs.

Development control was made effective by requiring all proposed developments to gain permission from the planning authority. Whilst there were some exceptions to this rule, and whilst there was a right of appeal against refusal, development control really became the central platform of local planning policies since there was a certainty attached to the grant or refusal of planning permission that did not exist in relation to development plans. Finally and, some would argue, most centrally, were the financial considerations necessary to carry out the intentions of the 1947 legislation:

In taking away from owners the legal right to develop their property without express permission, the Act quite logically also gave this legal restriction economic expression: it in effect simply expropriated the development value in all land, both that existing at the time, and all that development value which would be created in the future. If owners were given permission to develop their property, they were therefore obliged to pay a 'development charge', which was to be calculated as 100 per cent of the difference between the value of the land with this planning permission and its value without it. What all owners kept, was the existing use rights in their land; no land was to be allocated in plans for a less profitable use than the existing one. (Reade 1987, 20).

Having got this machinery in place, the Labour Party fully expected all transactions for development land to take place at existing use value since the excess would simply be taxed away. This did not, however, occur since the demand for such land rose so sharply that prices rose regardless of the effects of taxation. When the Conservatives were returned to power at the start of the 1950s they lost no time in declaring that the development charge was actually a tax on developers, and repealed it. However, in not introducing compensation for those subject to planning restrictions, the repeal of the development charge left the system in a state of disarray where development value appeared to belong to the individual if planning permission were forthcoming, but to the state if it were not:

The profits to be made in such a system are largely created by planning, a situation which is obviously socially unjust. But the public sector, in such a system, also creates problems for itself. The land which must be acquired for non-profit-making public purposes has itself had its prices pushed up by planning restrictions, and that taxpayers must pay these inflated prices is an added injustice. (Reade 1987, 22).

Because of the restrictions on development, much pressure was placed on the countryside, both on aesthetic and recreational grounds. This was addressed by the creation of National Parks and Areas of Outstanding Natural Beauty under the National Parks and Access to the Countryside Act 1949. By 1954 ten National Parks had been created, covering 10 per cent of the land area of England and Wales. Given this new provision it might have been understandable that local authorities felt able to accord a low priority to other forms of recreation in the new development plans. However, the location of the National Parks did not (and still does not) accord with the national distribution of population, with the closest Park to London and the south east being the Brecon Beacons, some 120 miles (195 km) west (see Figure 5.1 for the location of National Parks). Indeed, the selection of the areas for the National Parks reflected the predominance of conservation objectives, whilst the subsequent management of the Parks has maintained this bias.

It must, furthermore, be appreciated that even in the new National Parks there was no presumption in favour of public access or outdoor recreation. Indeed, the National Park boundaries did little more than signify the existence of a modified planning area with a presumption against development that was even stronger than in areas outside the parks. Since existing landownership rights were unaffected by the designation of the National Parks, these planning restrictions allowed even less opportunity for recreation development, meaning that whilst being national in name, the parks were still essentially private in nature. There have always been opportunities for public access to open country within the National Parks, however, but by permission of landowners such as the National Trust, the Forestry Commission, water authorities (water companies since 1990) and, occasionally, the local authority and some private individuals, rather than as of right.

Given the fragmented nature of public access, even in the National Parks, the concept of long distance footpaths (now known as National Trails; see Figure 5.1) gained in popularity. The National Parks Commission had been

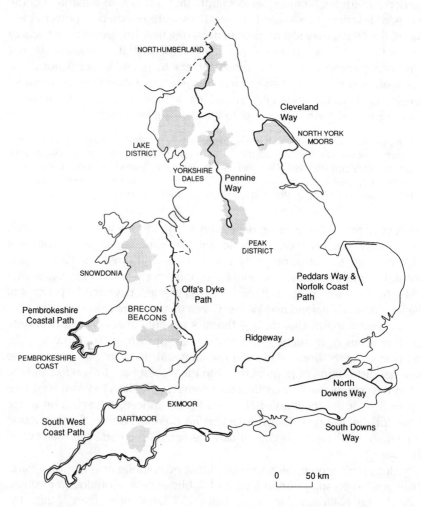

Figure 5.1   National Parks and National Trails in England and Wales

given the responsibility for their designation, in the National Parks and Access to the Countryside Act 1949, as a means of giving people the right to walk through the countryside without having to purchase or expropriate rights from private landowners. Although fraught with difficulties (Seeley 1973), preparation and designation of these paths proceeded apace, so that five major routes had been designated by 1955, including the Pennine Way, the Cornish Coastal Path and the Offa's Dyke Path. In the 40 years since the enactment of the National Parks and Access to the Countryside Act 1949 the designation of, and support for, national trails has become one of the central features of government policy on public access, carried out initially by the National Parks Commission and, since 1968, by its successor, the Countryside Commission.

Within the development plans for urban areas, the Town Maps and the Comprehensive Development Area Maps, land and sites were being allocated for recreation purposes. In particular, areas of public open space were being designated, usually in accordance with the National Playing Fields Association standards, whilst sites were also being designated for more specialist facilities. However, the priority given to these uses was low, largely because they were perceived by the local authorities as being less important than housing, industry or commercial uses of land. This was further exacerbated by the absence of the private sector from recreation development on the grounds that it was not commercially viable. This meant that there was little pressure from the private sector to provide or allow sites for recreation development, so confining such pressure to the narrowly defined role of the local authority.

Regardless of the removal of the financial considerations of the 1947 legislation, it lasted for two decades without fundamental review. During this time it was felt to have been a qualified success, but more for its role in development control than as an overall planning concept (Allison 1975). As far as the development plans were concerned, they were found to be too cumbersome and inflexible, meaning that they could not be kept sufficiently up-to-date (Cullingworth 1976). Indeed, it took until 1961 for all 145 planning authorities in England and Wales to get their plans approved, with a further seven years before the same process was complete in Scotland. Even after approval, however, the plans were still subject to interpretation by planning officers and committee, or by an inspector at appeal. This really meant that the legislation had been ineffective in securing positive planning, being mostly reliant on its control functions to direct the development intentions of the private sector.

In May 1964, in response to these criticisms the Ministry responsible for the planning system, the Ministry of Housing and Local Government, set up the Planning Advisory Group to review the system. The Group focussed its attention on the development plans, coming to the view that:

the main defect of the present planning system lies not in the methods of control but in the development plans on which they are based and which they are intended to implement. (Planning Advisory Group 1965, para 1.15)

After reviewing the operation of the development plan system the Group suggested that any replacement should be able to:

1 guide the urban development and renewal which is certain to take place;
2 promote efficiency and quality in the replanning of towns;
3 encourage better organisation and co-ordination of professional skills so that town and country are planned as a whole; and
4 stimulate more purposeful planning of rural and recreational areas.
                                (Planning Advisory Group 1965, para 1.33).

It is revealing that the Group identified rural and recreational planning as one of the four priority requirements for the development plan system:

There is no doubt that the explosive growth of population and car ownership, increased personal incomes and greater leisure will have a tremendous impact on country and coastal areas. This will be reflected in the demand for recreational facilities, access to the countryside, development pressure in villages and rural areas and intense traffic demand at weekends and holiday periods. These are among the most important problems that planning will have to cope with in the next twenty years and they are problems of the counties. (Planning Advisory Group 1965, para 1.33).

Following the report, which the Labour administration accepted fully, a number of measures were taken, including the introduction of the Land Commission Act 1967, whereby the state could buy land at current use value and then make it available to developers, and the Countryside Act 1968 which set up the Countryside Commission and allowed local authorities to provide a much expanded range of countryside recreation facilities. This was followed by reform of the 1947 legislation in the Town and Country Planning Act 1968 which introduced two tiers into the development plan system: the Structure Plan, to be prepared as a policy statement and approved by Westminster; and Local Plans, to be developed out of, and to give effect to, the Structure Plan. However, whilst the two tier development plan system brought some relief from the old style plans, the radical approach to development land envisaged in the Land Commission Act 1967 never proceeded very far as the newly formed Land Commission did not possess a sufficient budget to acquire city centre or other valuable land. In the event, the Land Commission was disbanded by the incoming Conservative government at the start of the 1970s. However, fuelled by the removal of the betterment tax and the Land Commission, land prices began to rise steeply, forcing even the recently-elected Conservatives to introduce a capital gains tax for development values in 1973.

On the basis of the views expressed in the Planning Advisory Group's report and the general endorsement of these by the government, a more positive approach towards recreational planning on the part of local authorities appeared to be likely in the revised development plan system unveiled in the Town and Country Planning Act 1968. This was further strengthened by the publication in 1970 of the Government's manual on the preparation of the new development plans (Ministry of Housing and Local Government 1970). Within this guidance on how to operate the new development plan requirements, recreation and leisure provision were identified as subjects to be included in the new county structure plans. In addition, the guidance also suggested that they might be suitable as a subject local plan, particularly if there was an attractive recreational feature in the area, such as a river valley or a strip of coast.

However, the promise of positive action was not fulfilled. The publication of the Development Plan Manual seemed to produce confusion rather than clarity in the minds of local planning officers. Most counties proceeded to cover a broad range of topics in their structure plans and some even went as far as attempting to develop their structure plans into corporate planning documents moving, in the process, away from the old land-use orientation that planning had previously espoused. In its place came a focus on socio-economic policies, so extending both the scope and content of town and country planning.

Rather than being a fault of the new system, however, the principal reason underlying this interpretation was the reform of local government under the Local Government Act 1972. With the introduction of two tiers of local government the planning process that had been designed for single tier operation had to be altered. This meant county councils having respon-sibility for structure plans and certain strategic development control func-tions such as minerals and highway issues, whilst district councils were given the general responsibility for local plans, to be derived from the structure plans, and most development control matters. This split between the tiers obviously caused much friction, especially when the county structure plan forced measures on district councils that they did not want to accept, and when the legitimate scope and content of structure and local plans was still the subject of much debate.

Because of these problems of power and interpretation at local level, central government reacted swiftly, for both practical and ideological reasons. The main practical problem was felt to be the excessive time that such a broad approach to structure plans would take to implement; one of the principal reasons for introducing the new system in the first place. The ideological reasons were more closely associated with what government felt should be the proper role and concern of planning. The central problem that the Department of the Environment had to address in terms of the scope and content of structure plans was well expressed by Solesbury, when he wrote that it was necessary:

> ... to decide to what extent the new Structure Plan is a broad social, economic and environmental policy vehicle as compared with a mainly physical development policy vehicle which took account of social and economic matters. (1975).

However, it was the Labour administration elected in 1974 that had to bear the consequences of the problems. This did not deter the new government from attempting to introduce what Reade (1987) has described as the most radical overhaul of the planning system since its original inception. With the enactment of the Community Land Act 1974 and the Development Land Tax Act 1976, Labour attempted to transform development control into a force for positive planning. Under the former legislation, all development land would have to be bought and sold by the government, so allowing it the ability to dictate which parcels of land would benefit from planning permission, and when. The latter piece of legislation was simply designed to ensure that the state was the recipient of any development value (that is, any increase in the sale value of land that resulted from the gaining of planning permission for a change of use) that would otherwise have accrued to the private sector.

Unfortunately for Labour, these measures were introduced at a time of recession and property slump and were, consequently, rendered largely irrelevant. Equally, since the legislation was to be imposed by county councils, most of which were firmly Conservative, there is some doubt as to how hard those councils tried to make it work. Following the Conservative victory in 1979 the Community Land Act was swiftly repealed, although the Development Land Tax Act was retained until 1985 as a partial means of funding local government.

In response to the initial problems with the new development plan system, the Department of the Environment decided to issue a series of Circulars to provide guidance to local planning authorities on the subjects to be addressed in their structure plans. In Circular 44/71 the view was expressed that, although the list of subject matters was potentially very wide, local authorities should in their surveys and structure plans select and emphasise those fields which were especially relevant to the problems and circum-stances of their particular area. By 1974 the official view had become much more definite. In Circular 98/74 the Department of the Environment instructed local authorities that in preparing their structure plans they were to consider only those issues thought to be of key structural importance. These were felt likely to be:

   a) the location and scale of employment;
   b) the location and scale of housing; and
   c) the transportation system.

Provision for recreation and tourism was, consequently, not felt to have key structural importance in any area, although both were included in a supplementary list of other issues that might be of particular importance in any given situation.

The understandable perception by local authorities of the low priority attached by central government to recreation was confirmed by Circular 55/77, a general memorandum on structure and local plans, which con-tained no reference to recreation at all. A similar circular two years later

(Circular 4/79) likewise ignored recreation, although Circular 13/79 did emphasise the importance of tourism, as distinct from recreation, to the national and local economy. When recreation eventually appeared in the next general memorandum (Circular 22/84) it was relegated to the last of the substantive issues to be considered.

The likelihood of local plans covering recreation issues was similarly small. Local Plans Advice Note 1/76 asserted that too many plans were being proposed, often of an inappropriate type. The Department of the Environment stated that:

> Local Plan coverage for areas of planning need should be aimed for, to enable development to be controlled effectively and economically and to provide an up-to-date planning context for the Community Land Scheme. (Local Plans Advice Note 1/76, para 2).

Local Plans Advice Note 1/78 further elaborated on what the Department of the Environment perceived to be the priorities for local plan preparation. These were: action areas defined in structure plans; areas where the use of the powers under the Community Land Act were envisaged to initiate development; areas with a large industrial component; areas with inner city problems; major growth areas indicated in the structure plan; areas under pressure for development; areas with land availability problems; and areas where particular planning issues needed to be resolved. Again there was no formal place for recreation within the local plan process. The same attitude appeared in Circular 23/81, referring to development plans in general, where the Department of the Environment stated that there should be a clear need before a local plan was prepared. It further stated that a local plan would not be required where the structure plan provided an adequate planning framework or where little or no pressure for development was expected.

On the basis of this clear and repeated guidance from central government as to the priority issues in structure and local plans, it is perhaps hardly surprising that the attention given to recreation has been strictly limited. The use of local subject plans has not been encouraged by the Department of the Environment, which is unfortunate since it would seem to provide an excellent means for the consideration and expression of recreation policies. There are examples in practice, such as the Berkshire Recreation Subject Plan, but all too few.

However, the very role of the development plan system has been in considerable doubt over the last ten years, with repeated attempts to reduce its influence on the development process. The intention behind the 1947 planning legislation was that whilst the development plan was only a guidance document it was to have a significant effect on decisions in development control. This did, in fact, occur in the period up to 1980, although the influence of the actual plans depended upon how recently they had been agreed and how relevant they were to the proposals under consideration. By the start of the 1980s, however, the emphasis of national policy was on economic regeneration, with local planning authorities exhorted, through Circular 22/80, to put as few restrictions on development as possible, by stating that:

> ... development is only to be prevented or restricted when this serves a clear planning purpose and the economic effects have been taken into account. (Department of the Environment Circular 22/80).

The Department of the Environment went on to make it plain to local planning authorities, through Circular 22/80, that development was not to be restricted by unjustified obstacles, especially if the proposals were for 'industry, commerce, housing or any other purpose relevant to the economic regeneration of the country'. Clearly the issue of planning for recreation was not seen as a high priority at this time, whilst even government support for tourism development was only as a result of its potential as a vehicle for economic regeneration (Circular 13/79).

The emphasis on national policy was further extended in Circular 14/85 when the role of development plans was significantly downgraded by requiring planning authorities to take a positive view of all applications unless there were strong grounds for rejection:

> There is therefore always a presumption in favour of allowing applications for development, having regard to all material considerations, unless that development would cause demonstrable harm to interests of acknowledged importance. (Department of the Environment Circular 14/85).

Not surprisingly, the debate as to what constitutes 'demonstrable harm to interests of acknowledged importance' continues to rage. Circular 14/85 took the government's perspective on the nature and operation of the planning system a stage further by undermining the role of locally generated policy. It did this by stating that development plans are one, but only one, of the material considerations that must be taken into account in dealing with planning applications:

> They (*development plans*) should not be regarded as overriding other material considerations, especially where the plan does not deal adequately with new types of development or is no longer relevant to today's needs and conditions – particularly the need to encourage employment and to provide the right conditions for economic growth. (Department of the Environment Circular 14/85).

Central government's opinion of the relevance and content of development plans was confirmed in the consultation paper 'The Future of Development Plans', which was published in September 1986, in which it proposed the abolition of structure plans, to be replaced by statements of planning policies drawn up by the counties. Whilst these statements were to cover a wide range of topics, recreation was not considered sufficiently important to be one of them. These proposals were followed, some three years later, by a White Paper of the same name. Once again the list of topics to be addressed did not include recreation. However, the White Paper went on to suggest that any extension beyond this 'core' of strategic issues, such as recreation, would require the prior agreement of the Secretary of State for the Environment.

More recent guidance from the Department of the Environment, directed by a new Secretary of State appointed in 1989, has reversed this trend. In the 1990 draft Planning Policy Guidance Note PPG 15, titled 'Regional Planning Guidance, Structure Plans and the Control of Development' it was stated that county structure plans would continue to be a key element in the development plan system. Not only did this effectively bring to an end a period of considerable uncertainty, it also introduced tourism, leisure and recreation as one of the key structure plan topics for the first time. Draft Planning Policy Guidance Note PPG 15 also renewed the role and significance of the development plan itself, by replacing the presumption in favour of development with a presumption that an up-to-date development plan would carry considerable weight in the determination of a planning application. Not only did it clarify the role, function and content of the plans, therefore, but it also gave the policies in the development plan greater priority as an influence on the determination of planning applications. This has recently been reaffirmed in the Planning and Compensation Act 1991, which requires planning applications to accord with the development plan unless material considerations indicate otherwise. The 1991 Act also introduced some important requirements with respect to recreation and amenity, firstly by requiring all National Park authorities to prepare park-wide local plans, in line with the requirement that all non-metropolitan districts should prepare district-wide local plans and secondly, that all these development plans should include policies in respect of the conservation of the natural beauty and amenity of the land. Quite what this second requirement will amount to remains to be seen; whilst it is undoubtedly focussed on conservation and the environment, any policies concerned with amenity must affect recreation and access to the countryside.

Despite these developments, however, there is relatively little evidence of significant shifts in policy direction with respect to recreation. Indeed, the only major references contained in the recent regional planning guidance have been, once again, to the potential economic impacts of tourism. Thus the Regional Guidance for East Anglia (Department of the Environment RPG 6) recommended positive support for tourism on account of job creation and the effects of visitor spending. Equally, the Regional Guidance for the South East (Department of the Environment RPG 9) recommended that:

> in making provision for development, priority should be given to industry and commerce, including tourism, particularly in growth sectors serving regional, national or international markets. (Department of the Environment RPG 9, para C.7).

Alongside the recent affirmation of the importance of tourism, leisure and recreation in the preparation of structure plans, Planning Policy Guidance Note PPG 12, issued at the end of 1988, contained similar reference with respect to local plans. The net effect of these two Planning Policy Guidance Notes is that:

> The structure plan may indicate in broad terms the areas in which facilities for recreation, leisure and tourism are to be provided and include general proposals for developments such as country parks, long distance footpaths and major indoor, outdoor and water sports facilities. Detailed development plans (*presumably local plans*) may, where appropriate, set out development control criteria, and identify sites ranging, for example, from a regional park to a neighbourhood play area, and policies for caravans and camping. (Department of the Environment Planning Policy Guidance Note PPG 15).

This growing recognition, on the part of the Department of the Environment, of the importance of recreation provision has recently been further confirmed by the publication of a Planning Policy Guidance Note on Sport and Recreation in September 1991. However, whilst it is undoubtedly significant that such guidance has been produced, the quality of the advice is debatable, with little apparent recognition of the potential future role of recreation in society, nor of the problems facing local authorities in balancing future provision between the public and private sectors of the economy.

The last decade has seen two opposing forces, with the government ideologically opposed to planning in its bureaucratic sense, but more than willing to use it as a means of industrial relocation and regional development, as well as a means of protecting landowners and, more generally, the property industry. Thus the attack on planning that has occurred could be largely symbolic, aimed at undermining the concept of 'planning' and state control, rather than directly at the planning system itself (Reade 1987). However, what is equally clear is that the elements of development control that have been the cornerstone of Conservative planning ideology have not worked appreciably better than any other part of the system:

> My question, then, was whether the new planning techniques of the 1960s and 1970s have led to any improvement in development control. The answer would seem to be that they have not. There has been no significant improvement, since 1960, in the planners' ability to assess the consequence of their actions. The 'new' planning techniques, even when of a 'substantive' rather than 'procedural' nature, have been applied almost entirely to plan-making rather than to discovery of the effects of planning, and as a result there has been very little, if any, growth in the knowledge-base of planning. (Reade 1987, 109–110).

In conclusion, therefore, it appears that a system that was first formulated to deal with a specific national problem has subsequently declined into a largely reactive process with little focus or basis in positive planning. This decline is particularly pertinent for recreation planning since the provision of open space for recreation and leisure was such a central part of the original reasons for introducing a statutory planning system. Since that time the importance attached to recreation provision in the planning process has suffered to the extent that it has become no more than a residual consideration once other land uses have been accounted for.

5.4 Conclusion: the place of recreation provision in the statutory planning system

As this chapter has indicated, recreation has always had a low priority in the statutory planning process, regardless of the fact that one of the earliest reasons for promoting town planning was the loss of open space as it was sold for residential and industrial development. Indeed, most of the forces that have worked to shape the modern planning system have also, largely inadvertently, served to marginalise the place of recreation within that system. In the countryside the general neglect of rural planning, with its inherent bias towards agriculture as the primary use of land, has ensured that recreation has only gained in importance as the income from agriculture has declined. Equally, when the value of a particular use such as recreation is largely social and consequently not recognised in the market, it is destined to be no more than a residual use of land that has no other value, whether in the town or country. The only method of ensuring that this social value is taken into account, therefore, is either for the state to purchase land for that use, regardless of its price, or to intervene in the market by removing the incentive to sell to the highest bidder.

Indeed, the whole framework upon which planning has been predicated has, for the most part, tended to neglect recreation. By largely basing development plans on land use zoning, it has tended to subjugate multiple uses in favour of primary ones, and future uses in favour of the past and present. This means that even in areas where provision for recreation is seen as important, such as the National Parks, primary uses such as agriculture and forestry still dominate, whilst the inevitable balance that must be struck between recreation and conservation is rarely achieved through the genuine multiple use of land, but rather through the compromise of zoning. Equally, the reactive nature of the planning process means that opportunities to secure recreation provision are not often taken up. This is particularly so in the countryside where, for example, permissions for the extraction of minerals are still often accompanied by a requirement to restore to the former use, rather than seeking a more innovative future use, such as recreation.

A notable exception to this view of the planning system as purely reactive has been the long-running debate over the extent to which the system has created the market for development land and whether and how society should benefit from this. As the demand for urban land began to grow in Victorian and Edwardian times, consequently pushing up land prices, the dominant view was that individual land owners should not gain from a need for land on the part of society; that any increase in value should be the property of the society that created it. Whilst this view prevailed it was possible to imagine a system of positive planning where land uses were determined according to the problems faced by society rather than by the operation of the free market, consequently providing a role for recreation and open space. However, this situation was never likely to be favoured by developers or the owners of land in cities since it deprived them of the chance to make windfall profits as well as meaning that the future use of their land could be determined bureaucratically.

For these reasons repeated attempts to separate the ownership of land from the development value of that land have failed, even in as far as any attempt at formalising a system of betterment tax has been concerned. This has meant, since 1953 at least, that the anomalous situation has existed where owners are free to make whatever gains the planning system allows or facilitates, but where any restriction of that value by the planning system does not lead to compensation being paid by the state. This has inevitably put an ever-increasing pressure on the planning system to grant permissions, even if the proposed land use is not considered optimal for the site, since it has become a 'winner takes all' situation. When major proposals are backed by the finances necessary for an appeal and public inquiry, the pressure on the planning authority is further intensified.

Quite understandably, the outcome of this situation has been the opposite to the positive planning upon which the system was originally conceived. Rather than development plan guidance backed up with the power of control, the system has become no more than a bureaucratic tool to be used at the convenience of government:

> Government may in fact be using the DoE (*Department of the Environment*) and the planning system to do for the property industry what political scientists have long seen the Ministry of Agriculture as doing for the farming interest – that is, to strengthen its hand, to provide the conditions in which it can operate profitably. (Reade 1987, 65).

Since the abolition of the Development Land Tax in 1985, therefore, it has been implicit in the planning system that all development value is attached to the land, however it is created. Rather than accepting that the land market would be the sole means of determining the use of land, however, planning authorities have continued to try and extract a form of betterment. This has been through the medium of planning agreements, where developers agree to provide certain facilities, or to provide for certain uses, within their proposed scheme. This system, known euphemistically as 'planning gain', has been directly responsible for the provision of many leisure facilities that would otherwise not have been developed. In reality, however, the system is much more akin to developers using the development value of their land to 'buy' planning permissions; to privatise the betterment concept.

Whilst originally seen as a potential way of modifying the market mechanism enough to gain desirable community facilities without questioning the ownership of development value, planning gain has now got to the point where it is being used openly and blatantly to influence the planning authority. A recent case in point is an application by a major oil company to build an artificial island for oil extraction in Poole Bay. Having initially been against the scheme, the councils of Bournemouth and Poole have apparently been persuaded to grant permission on the basis of a £1m payment towards the development of the region's tourist facilities. Whilst there has never been any suggestion that the councils were in any way entitled to this money, it is in reality a betterment payment for the right to develop, regardless of the undesirability of that development or the relative social value of not developing the area.

Because of this situation it is clear that many uses of land, including much recreation, are not properly accounted for in the planning system. Indeed, few recreation facilities can compete on a commercial basis for land in urban areas, meaning that such facilities are constrained to sites already owned by the state, or provided as part of a planning deal, usually in association with other types of development. At the very least this means that the recreational 'needs' of society cannot be planned in a coherent sense, regardless of which sector of the economy will ultimately be responsible for provision. More fundamentally, however, it means that recreation, like a number of other 'peripheral' land uses, can be blatantly used to achieve political ends. Whether in terms of the concern to protect open space expressed by Nineteenth century industrialists or the interest in tourism expressed by central government at the end of the 1970s, the provision of land for recreation has been manipulated to help achieve the wider goals of those in power.

The following two chapters will now explore the relationship between policies for recreation and the actual provision of facilities, firstly in rural areas (Chapter 6) and then in urban areas (Chapter 7).

6 Rural Recreation and Farm Diversification: Control of Development in the Countryside

6.1 Introduction

This chapter will seek to trace the development of rural recreation and farm diversification in an attempt to illustrate the way in which the countryside has been at the centre of some of the most fierce political battles over the development of recreation as well as, more recently, the scene of some of the most radical policy shifts by government. Unlike many aspects of recreation provision, outdoor recreation has never been constrained by the availability of facilities in the same way that indoor sport and recreation have. Indeed, in initially reviewing the demand for outdoor recreation it is apparent that it has traditionally relied on publicly available resources such as rights of way and, to a lesser extent, commons. This has not prevented people from demanding increased rights of access to the countryside. However, whilst this demand has partially been met through the provision of publicly and quasi-publicly owned land and facilities, as in urban areas, a political struggle has continued over the rights of individual landowners to exclude the public from walking over their land (Shoard 1987 provides a full account of this struggle).

This protection afforded to landowners has, in fact, remained one of the central facets of rural recreation policy for over a century now. This has given landowners an opportunity when considering the supply of recreation facilities in the countryside. Since the enactment of the National Parks and Access to the Countryside Act 1949, landowners have been able to enter into access agreements with, principally, local authorities. These agreements have allowed public access to privately owned land in return for appropriate consideration, either through an annual fee paid by the local authority, or through free wardening services or other such arrangements. Equally, many landowners have opened their parks, stately homes and gardens to the paying public, secure in the knowledge that people will pay for the privilege of getting into the countryside.

Until fairly recently the majority of formal recreation development in the countryside was provided either by the public sector or by the owners of estates and large land holdings. Farmers, as managers of the vast majority of the countryside, took little part in recreation provision, concentrating mainly on food production. The only real exception to this was the incidence of farm-based accommodation in the traditional tourist areas of Wales and the West Country of England. However, with the rising cost of subsidy and over-production of many foodstuffs, recent agricultural policy has featured a squeeze on production and, consequently, farm incomes. This has co-incided with a general policy shift away from agricultural fundamentalism towards a more balanced view of the countryside. This balance has meant that the demands of a largely urban public are being given a higher priority whilst farmers and landowners are being given the support to provide the facilities demanded by those people. Thus a policy that was designed to constrain development in the countryside in order to protect the land of the wealthy, the incomes of the farmers and the production of food for the nation has been replaced by one that is, if not encouraging then facilitating, development in the countryside in order to continue to protect the ultimate tenure of the land of the wealthy and the incomes of the farmers.

6.2 Outdoor recreation and the demand for access

Whilst rural recreation has existed as long as the concept of leisure and recreation, the modern advent of rural recreation can be traced to the industrial revolution in Nineteenth century Britain and the movement of working people from the land into the towns and cities (Joad 1945). Indeed, in the century to 1939 the proportion of those living in urban areas grew from one-third to two-thirds of the population of England. Whilst people may have been consigned to live close to their work, substantial numbers of them (of both working and middle classes) did attempt to walk in the country during the free time available to them.

Thus it was actually townsfolk who began to demand greater public rights of access to the countryside and who, in 1865, formed the Commons, Footpaths and Open Spaces Preservation Society to fight the intended enclosure of some commons and open spaces. This society later led to the formation of the National Trust in 1895, to further the preservation of the Victorian commons for public recreation (Blunden and Curry 1989). The last quarter of the Nineteenth century also saw the first formal attempts to gain a legal right of access to moors and mountains. Following various con-frontations with landowners and gamekeepers, particularly in Scotland, James Bryce introduced the Access to Mountains (Scotland) Bill to Parliament in 1884, with the intention of gaining a public right of access to all moors, mountains and open land in Scotland. Although the Bill failed, as did a number of others that followed, there is evidence that the access movement had some support from all the political parties at this time (Hill 1980).

The introduction of mass public transport, called the second industrial revolution by Joad (1945), heralded a great increase in the demand for countryside recreation and with it, an increase in the pressure for public access and the incidence of conflict between walkers and landowners. This culminated in the 1930s with the formation of the Rambler's Association as a united political movement intent on gaining greater public rights of access. Other groups were also formed, on both sides of the political spectrum. None of these was making progress, however, and the goal of public access, the subject of a further 16 unsuccessful attempts at gaining legislation up to 1931, was slipping further and further from their grasp (Hill 1980).

In desperation of ever making any progress, one of the more militant groups, the Lancashire branch of the British Workers' Sports Federation, decided upon direct action. After a small-scale trespass over private land in the early 1930s, the Federation organised a group of ramblers to storm one of Britain's most inaccessible grouse moors – Kinder Scout in the Peak District – a moor that had been enclosed by the owner during the previous century. The mass trespass happened in April 1932, with around 800 ramblers taking part. Although generally peaceful, six people were arrested, five of whom were members of the Young Communist League. Despite minor charges such as unlawful assembly and breach of the peace, a jury comprising 11 landowners found them guilty and they were jailed for up to four months each. Other trespasses followed, confirming the conviction of the ramblers, although it is doubtful how much support they engendered.

Almost concurrent with the trespass movement was the work of the Addison Committee, charged by Ramsay MacDonald in 1929 with considering the establishment of national parks, both as a conservation and a recreation measure. Addison's (1931) report suggested a hierarchy of designated areas, with the most highly protected being conservation areas. The concept of public recreation areas also found favour with the government. However, the ensuing recession and, eventually, preparation for conflict ensured that the proposals got no further at that time (Chessell 1945).

The aftermath of the mass trespass saw many influential people join the access movement and provide it with a hitherto unknown acceptability. In recognition of the fact that the public were beginning to favour public access (Blunden and Curry 1989), the landowners in government appeared to have something of a change of heart, so allowing a private member's bill to come before Parliament in 1939. However, what started as an access charter was wilfully altered in the committee stages of the Bill to become a landowner's charter, making trespass a criminal offence for the first time since the reign of Charles II. Although the trespass clause was subsequently dropped, and the whole Access to Mountains Act 1939 never really acted upon, the way in which the Bill was altered served as a timely reminder of the continuing power of the landowning lobby in Parliament.

Having been in favour of the Bill on its passage through Parliament, the access lobby was then forced to oppose it and seek to have it repealed (Stephenson 1987). However, the war intervened. Immediately after the war the pressure again mounted on the government to alter its stance on access. Amongst the protagonists was the philosopher C.E.M. Joad, who argued that the new-found prosperity of the nation was going to put an intolerable pressure on the countryside which could only be realistically relieved by allowing greater freedom of access to the general public. (Joad 1945).

The result of the lobbying, and of a report commissioned by the government from Scott (1942) on land utilisation in rural areas, was the recognition that access to the countryside could not be denied for ever. This led to John Dower, then a temporary civil servant, being commissioned in 1942 to produce a report on national parks in England and Wales. His report (Dower 1945) included proposed areas for the parks, although it did not go as far as Scott (1942) in advocating a central authority to oversee the development of public access. Indeed, in not following the lead provided by Scott (1942), Dower inadvertently supported the claims of the private landowners and farmers that they were best placed to look after the countryside for the Nation. In particular, Dower's (1945) arguments rested on two important assumptions that were to underpin the implementation of national park policy. As MacEwen and MacEwen explain:

> One was the idea that the town and country planning system (which did not then exist except on paper) would reconcile private landownership and the exploitation of the natural resources of the parks with the public good and ensure a harmonious, well-ordered, well-designed countryside. The other was that a prosperous farming industry would preserve both the rural landscape and the rural communities. (1982, 9).

After the war the new Labour administration appointed Sir Arthur Hobhouse to look into Dower's work and produce a working plan for establishing the national parks (Hobhouse 1947). The result of these committees was the National Parks and Access to the Countryside Act 1949. It was seen at the time by Lewis Silkin, then Minister of Town and Country Planning as:

> A people's charter – a people's charter for the open air, for the hikers and the ramblers, for everyone who loves to get out into the open air and enjoy the countryside. Without it they are fettered, deprived of their powers of access and facilities needed to make holidays enjoyable. With it the countryside is theirs to preserve, to cherish, to enjoy and to make their own. (quoted in Blunden and Curry 1989, 63–4).

However, the reality was somewhat different, for the statute enshrined the pre-war policy stance of the government and became the basis for all policy that has followed over the past 40 years. Whilst setting up a National Parks Commission and designating national parks (see Figure 5.1), the Act was really based on retaining and, indeed, enhancing the primacy of private property rights. Neither in the national parks nor on any open countryside was there to be a general right of public access, far less any concept of nationally owned land. Furthermore, if a particular landowner was minded to allow access, it could be done via an access agreement made with the local authority, in return for compensation for the disturbance and possible damage caused.

Whilst these new access agreements did meet with some success, principally in the new Peak District National Park which had been the subject of much previous conflict over walkers' rights (along with conflicts over issues such as quarrying and the siting of power lines), most access continued to be either via the rights of way system or on publicly or quasi-publicly owned land, as was the case before the Act. Primarily, this public land belonged to the Forestry Commission, the water boards and the local authorities, with the National Trust providing a fast expanding resource of quasi-public land, acquired in the name of the state in lieu of death taxes.

In addition to its role in dealing with access to the countryside, the 1949 Act also comprised the State's policy for conservation. This part of the Act appealed in particular to the landowning lobby, since it tended to legitimate their claim to custodianship of the land. It was to be expected, therefore, that in reality recreation would play second fiddle to conservation, a situation that has remained unchallenged ever since.

Following the Second World War, the access lobby became split between a fast increasing number of non-political people who simply wanted to walk and a political minority who were still fighting for public rights. The expansion of the former category was phenomenal, with the Youth Hostels Association growing at the staggering rate of 1200 new members a day in the immediate post war period. The activities and growth of the political lobby, on the other hand, were somewhat curtailed by the 1949 Act and the involvement in agreeing the National Parks and seeing how the new access agreements would work.

Pressure for public access continued, however, with a switch towards the protection of existing rights, such as those over many commons. This led, in 1958, to the establishment of a Royal Commission on Common Land. Amongst other things, the Royal Commission advocated a system of registration for commons, formalised plans for the management of registered commons and a public right of access for quiet enjoyment and recreation. The Commons Registration Act 1965 fulfilled the first of these recommendations, but the latter two are still being debated three decades later (Clayden 1985).

The political impasse over rural recreation remained relatively constant over the next decade or so, although it became increasingly apparent that the mobility of urban people was bringing a new type of visitor to the countryside. This new visitor was the car-borne recreator. Rather less concerned with access to wild country, the new recreators did not wish to travel far from the towns and certainly did not wish to become separated from their cars (Seeley 1973). Whilst the government appeared not to appreciate the potential effect of this shift in emphasis, organisations such as the Civic Trust likened it to a new social revolution; the fourth wave sweeping over Britain (Dower 1967). However, it was not until 1968 that the government acted, although even then the underlying policy of private ownership remained unchallenged. Under the Countryside Act 1968, The National Parks Commission was replaced by individual park planning boards controlled by the local authorities within whose borders the national parks fell. At the national level, the Countryside Commission was created as a new advisory and grant aiding body for government. It was charged with overseeing recreation and conservation interests in the countryside as well as advising government on future policy. The 1968 Act also widened the power of local authorities to develop and manage recreation facilities in the countryside. Under sections 7 to 10 of the Act, they were empowered to provide facilities and carry out works, such as the construction of buildings or the provision of car parks, in country parks, picnic sites, camping grounds and on common land. These functions could not only be exercised on land owned by the local authorities, but on other privately owned land, subject to suitable terms being agreed with the owner. A local authority could also compulsorily acquire land in order to carry out these functions. Because of the broad nature of these provisions, the Act was implemented in very different ways in different parts of the country; in Berkshire, for example, the county council saw its primary function as being one of providing advice and co-operating with others, whilst Cheshire and Hampshire County Councils set up specialist recreation departments in order to effectively discharge their responsibilities (Miles and Seabrooke 1977). As Miles and Seabrooke go on to state:

> On the whole it seems that counties beyond the pull of London but with large towns or cities within easy reach of their countryside, have tended to set up separate and more or less autonomous countryside leisure departments and have taken to themselves the main responsibilities for recreation. In some cases, however, such leisure provision ... has been tucked in under the wing of existing county departments (1977, 37).

Whilst the shift from special areas such as the national parks to an overview of the countryside as a whole was to be welcomed, there was no actual change in policy. What had begun as protection of the landowners now became an obsession with containing the expected explosion in countryside recreation and resultant deterioration in the environment that was due to occur in the late 1960s and early 1970s. This policy of containment was enshrined in the new development control measures brought in by the Town and Country Planning Act 1971, whereby county councils could contain development in the countryside through structure plan policies.

The early 1970s saw a marked increase in the provision, mostly by local authorities and some owners of large estates, of country parks and near-urban destinations for a mass of visitors. These new developments were 'facilitated' by the Countryside Commission through advice and substantial grant aid. However, these new destinations were, for the most part, only available to those with private transport. This meant that countryside recreation policy began to shift away from the Labour-voting working class ramblers of the early decades of the century to being a preserve of a largely conservative, property owning, middle class. At the same time as there was a progressive democratisation of urban recreation, therefore, there was a concurrent withdrawal from democracy in the countryside.

Even the efforts of many local authorities, in providing new facilities and starting public transport services to the countryside at the weekends, could not arrest the growing alienation of the landless by the middle classes. This alienation was virtually complete by the mid-1970s when social survey work began to suggest that, on the one hand, few members of the working class were actually interested in countryside recreation (Blunden and Curry 1988), whilst on the other, those that did tended to take on the values of the middle class, both by spurning subsidised public transport and by being more than willing to pay entry fees to recreation facilities (Bovaird *et al*. 1984). By the start of the 1980s most of the subsidised transport schemes had been abandoned as failures, and with them most attempts to derive a social policy for countryside recreation.

Perhaps the most disturbing aspect of this period of recreation policy was the universal acceptance by the Countryside Commission and, to a lesser extent, local authorities, that the recreation patterns of the workforce had changed so fundamentally between the 1930s and the 1960s. That all those wanting access to the countryside in the interwar period had been replaced by a new urban workforce that had no interest in leaving the safety of the city, except, perhaps, to the odd safari park or on an organised coach tour. However, as Blunden and Curry (1988) discovered, the provision of certain types of facility and the support for subsidised transport were based on very little research and, as it transpired, little understanding of people's wishes. Perhaps most importantly, it must be recognised that both the Countryside Commission and county councils were really no more than voices of the landowning classes; indeed they remain so today. Thus government was able to maintain its policy of protecting landownership, but apparently at arm's length. Equally, the Countryside Commission saw its role very much as a facilitator, under the strict definition of the Countryside Act 1968. So whilst it might have been espousing the social conscience of government, it really had very little intention of encouraging the masses into the countryside. The Commission therefore undertook little relevant research, encouraged facilities and transport that it thought people might like, or which might be good for them, but was very quick to condemn them when they did not make good use of the opportunities.

However, by the early 1980s it was becoming apparent that the expected explosion in countryside recreation had not fully materialised. A combination of escalating oil prices, elitist land use policies and a rediscovery of urban living had combined to curtail the increasing demand for passive countryside recreation, whilst there appeared to be an increasing minority interested in more active pursuits, often involving sport more than recreation. Even the Countryside Commission began to recognise the failure of its early policies. In response, it moved away from the support of formal recreation facilities, such as country parks, to the promotion of informal activities and facilities. This particularly involved renewed interest in the designation of national trails, a general support for rights of way and the designation of areas of open country where people could enjoy 'wilderness' experiences. It also set up the Common Land Forum in the early 1980s and, when the Forum reiterated the findings of the 1958 Royal Commission (Common Land Forum 1986), lobbied government for a public right of access to all common land. It is immediately apparent that none of these forms of access was actually new or innovative. However, in renewing interest in them, the Countryside Commission did ensure that more people became aware of their existence. Equally, in tackling some of the problems of the designation of national trails (Seeley 1973), as well as taking a greater interest in the management and maintenance of all rights of way (Countryside Commission 1987A, 1989A and 1991B), the Commission did ensure that any increase in demand could be met by improved supply (see Figure 5.1 for details of the national trails and Table 6.1 for a list of their lengths and dates of designation).

Table 6.1 The designation of national trails

National Trail	Approved		Opened		Length (km)
Pennine Way	Jul	1951	Apr	1965	402
Cleveland Way	Feb	1965	May	1969	150
Pembrokeshire Coast Path	Jul	1953	May	1970	299
Offa's Dyke Path	Oct	1963	Jul	1971	270
South Downs Way	Mar	1963	Jul	1972	129
South West Coast Path:					
North Cornwall	Apr	1952	May	1973	431
South Cornwall	Jan	1954	May	1973	431
South Devon	Jun	1959	Sept	1974	155
Somerset and North Devon	Jan	1961	May	1978	168
Dorset	Apr	1963	Sept	1974	116
Ridgeway	Jul	1972	Sept	1973	177
North Downs Way	Jul	1969	Sept	1978	227
Wolds Way	Jul	1977	Oct	1982	127
Peddars Way and					
Norfolk Coast Path	Oct	1982	Jul	1986	150
				Total:	2761

Note: figures as at 31 March 1989
Source: Countryside Commission (1989B)

This period also saw an increase in the activity of the access lobby, perhaps in response to the avowedly free market government elected in 1979. In the first six years of Mrs Thatcher's leadership there were six private member's bills on access, some relating to all land and some to the commons alone. After a particularly intense period in 1983, when the Access to Commons and Open Country Bill and the subsequent Walkers (Access to the Countryside) Bill found some support in Parliament (although neither got close to being enacted), the government did agree to implement the recommendations of the Common Land Forum; something that the access lobby is still waiting for them to honour.

The mid- and late-1980s were characterised by a fundamental change in the balance of power in the countryside. However, this was not the long-awaited shift of power from landowners to the landless, but a remarkable attack on the dominance of agriculture as the cornerstone of countryside policy. On the one hand there were growing attempts to force landowners and farmers to carry out their duties with respect to current rights of way (Joint Centre for Land Development Studies 1985 and Countryside Commission 1988), whilst on the other there were attempts to reduce food production through, amongst other initiatives, the introduction of voluntary set-aside schemes designed to encourage farmers to reduce the amount of land under cultivation. More recently, the Countryside Commission has commenced its Countryside Stewardship Scheme (Countryside Commission 1991C) with the express intention of offering incentives to conserve, enhance or re-create certain types of landscape. Based on a system of capital and annual payments for conservation works, the Countryside Stewardship Scheme also introduced the concept of paying farmers and landowners to 'manage' public access, for the sum of £50 per hectare per year in 1991 prices. Having been well received by farmers and landowners, these initiatives have led, seemingly, to a rapid endorsement of non-agricultural operations and development on farms, both by the farming community and by policy makers.

Notwithstanding the introduction of paid public access within the Countryside Stewardship Scheme, the Countryside Commission, in its reassessment of the role of the countryside, focussed particularly on public access. In a striking reversal of much former policy it suggested that the time had come to reconsider the issue of unimpeded access to open country:

> Neither an unqualified 'right to roam', nor a policy of 'public keep out' matches the needs of the 1990s. Rather, we believe that there should be a presumption that in open country (primarily uncultivated, unenclosed areas) people should be able to enjoy quiet recreation on foot, unless they can be shown by their presence to be causing significant detriment to the other uses to which the land is put. (Countryside Commission 1991, 7).

Without actually seeking to undermine a landowner's or occupier's right of exclusive possession, therefore, this proposal would result in the onus being placed squarely upon landowners or occupiers to prove why access should not occur on their land. Predictably, this led to a swift response from landowners, via the Country Landowners Association when, in a recent publication on recreation and access to the countryside, it stated that:

> The CLA recommends that calls for an unqualified 'right to roam' should be resisted by the Government. Nor should the onus be placed on owners to show why public access should not be provided. Where there is a genuine need to extend public access to open country, local authorities should instead use their existing powers to negotiate voluntary agreements with owners and occupiers to

secure managed access, taking account of all interests. (Country Landowners Association 1991, 14).

The situation and likely future outcome of this renewed debate is far from clear; even the Countryside Commission recognises that the administrative machinery necessary to implement its proposals could not realistically be in place until the turn of the Century. However, it does seem clear that a renewed debate over peoples' rights of access to the countryside will occur and that, with the parlous state of agriculture and the consequent level of non-farming subsidies, the power of farmers and landowners may be insufficient to resist an extension of access rights.

6.3 The supply of farm-based recreation opportunities

The last ten years has seen a very confused situation in countryside policy. The start of the decade heralded the Wildlife and Countryside Act 1981 and the beginnings of environmental awareness in Britain. It also heralded the first concerted attack on the agricultural fundamentalism that has always driven countryside policy. These factors led, on the one hand, to the Countryside Policy Review Panel (sponsored by the Countryside Commission) calling for countryside recreation to be given equal status and funding with agriculture and forestry, whilst on the other, the Countryside Commission, on becoming a fully independent Grant-in-aid body in 1982, re-defining its role by increasing its commitment to conservation (Countryside Commission 1982), as if this had not always been the case.

However, political and financial pressures were the first to have an effect. With a major reform of the European Community's Common Agricultural Policy underway by 1982, reducing support prices and introducing levies and quotas, the spotlight was turned on helping farmers replace agricultural income with money from other sources. Recreation was soon seen as one such source, with the government's Farm Diversification Grant Scheme and a number of Department of the Environment planning circulars encouraging the development of non-agricultural enterprises on farms. Indeed, in the Agriculture Act 1986, the Minister of Agriculture was actually charged with the responsibility of promoting the public's enjoyment of the countryside.

With the consequent levels of interest and exposure given to farm diversification and the development of alternative enterprises, even informed bystanders might have been forgiven for believing it to be a new concept. It is, however, neither a modern phenomenon nor a new solution to problems that have been apparent periodically in the British farming industry. The current level of interest in farm diversification, especially within the farming community and its advisers and consultants, has tended to obscure the fundamental problem of how to maintain, or improve, income from the land (Byrne and Ravenscroft 1989). Indeed, the problem of income generation has rarely led to diversification, at least at the first instance. Rather, it has initiated a quest for improved efficiency in agricultural production (Paice 1988).

Diversification of the farm business has tended to occur only where no farming options have appeared to remain, where other members of the farming family have developed new enterprises in isolation from the farming business, or where the farmer has found a non-farming opportunity to increase income from the business. The latter of these reasons accounts for much of the early activity in farm diversification. Reports by Davies (1971, 1973) and the National Farmers' Union (1973) indicate considerable levels of diversification, particularly into farm-based accommodation and tourism, in the traditional holiday centres of Wales and the West Country of England by the late 1960s. Indeed, in a paper delivered at the Oxford Farming Conference in 1968, Wibberley suggested that the late 1960s were the beginning of a phase where many farmers could take advantage of the growing demand for recreation and tourism by diversifying into tourist provision (quoted in Davies 1971). This view has been subsequently borne out in work by Halliday (1989), who noted the long-standing nature of much of the farm-based accommodation in her 1986 survey of small dairy farms in Devon.

Since the late 1960s and early 1970s, the growth in farm diversification has been steady, with up to 30 per cent of all farms now including at least one alternative enterprise. This scale of development has been significant enough to precipitate a change in the way that the prosperity of farming is measured by the government. As Jones states, writing about the 1988 Review of Agriculture:

> In the latest national survey of farm incomes, it is for the first time inferred that the picture of farming's prosperity is not complete without a consideration of farmers' other sources of income. In particular it is argued that the scale of public financial support for farming should take account of farmer's total income and not just that obtained from farming, given that in recent years around 40% of the taxable income is estimated to have come from non-farming sources. (1989, 28).

A number of authors have produced lists or suggestions of enterprises that might qualify as diversification. In particular, Gretton (1985), Carruthers (1986), Parker (1987) and Slee (1987) all provide such information. However, there has been little information on the appraisal of such developments beyond the monthly bulletin *Farm Development Review* and any advice that can be obtained from sources such as the Ministry of Agriculture, Fisheries and Food's advisory service ADAS. Inevitably this has tended to mean that diversification has not been viewed in the same light as farming, even to the point where it is often felt to be quite separate from the farming business (Halliday 1989).

In terms of development on farms, farmers and landowners are essentially faced with four possible courses of action: to change or intensify the use of current assets; to convert assets; to exchange them for different ones; or to consider whether or not to engage in trading ventures as well as, or instead of letting assets to third parties (see Figure 6.1).

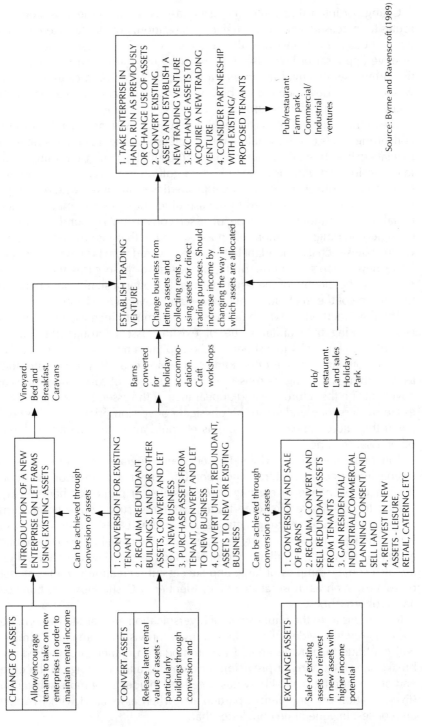

Figure 6.1   Diversification options for farms and rural estates

Source: Byrne and Ravenscroft (1989)

Change or intensification of the use of assets is one of the most common routes to increased income through diversification, both for farm tenants and owner-occupiers. It involves the retention of existing assets, but with a positive appraisal of how else they might be used in increasing income to the enterprise, or rental income to a landlord. A common example of this would be the development of bed and breakfast accommodation in the unused bedrooms of a farmhouse.

The farmer's second choice in seeking new sources of income is to consider converting assets to release latent rental values. In general, conversion tends to relate to the physical attributes of land and buildings, such as turning a barn into a residential unit, or parkland into a golf course. However, it can also relate to the conversion of the legal properties of an asset, where different forms of tenancy can be used to release different types and magnitudes of rental value. Although this type of asset conversion is found on owner-occupied holdings, it is more often associated with estate-owned let farms, where assets such as redundant barns are recovered from farming tenants and let to new tenants for non-farming purposes. In these cases there is little change in the income of the tenant farmer, but often a significant change in the income of the freehold owner (Byrne and Ravenscroft 1989).

The third method by which farmers can attempt to increase their income is by realising the capital value of existing assets through the sale of freeholds, followed by investment of the capital in new assets with a higher income potential. There is often some conversion of assets prior to sale, such as obtaining planning permission for a change of use and imposing covenants on the future use and appearance of the asset. Some farmers and estate owners may undertake the physical conversion prior to sale, especially if the future appearance of the asset could affect the amenity value of other assets retained.

The final method by which farmers can generate new income is by establishing trading ventures, incorporating either a change of use, conversion or exchange of existing assets. Estates in particular have always tended to be involved in some trading ventures, such as the home farm, forestry and some sporting activities. In many cases these have been supplemented more recently by the opening of the main house to the public, as well as providing ancillary enterprises, such as catering and retail. There are also some examples of large and successful recreation ventures started by estates, such as those at Beaulieu, Blenheim, Castle Howard, Longleat and Woburn Abbey.

Indications as to the future scale of diversification are hard to discern, but there is growing evidence of a dichotomy in provision. Many farm families will continue to provide limited accommodation, either through bed and breakfast, self catering or provision for caravans and camping. At the other end of the farming scale, Ilbery has noted the incidence of more substantial diversification on large farms, particularly when associated with livestock enterprises. In general he concludes that:

... recent changes in Government policy, with a relaxation of rural planning controls and encouragement of afforestation, farm woodland, golf courses, 'horsiculture' and small scale industry, should ensure a steady rate of growth. Indeed, with the twin problems of over-production and declining farm incomes, farmers will increasingly be forced to seriously consider a redeployment of resources to alternative enterprises. (Ilbery 1987, 34).

However, just as the farming industry was beginning to see this demand-led move towards diversification as a means of financial salvation, national economic recession has led to a severe downturn in the development industry, meaning that the high prices paid for small parcels of development land on farms has now passed, along with many of the more speculative aspects of the leisure market. In a recent conference, Sebag-Montefiore (1991) suggested that whilst there was still some scope for leisure development on farms, the tremendous golf course boom of the late 1980s had been overdone and there was increasing evidence that the banks were becoming reticent about lending money for leisure development projects. Seemingly, therefore, a trend that started out of a necessity to support farm incomes, and ran counter to the long-standing policy of containing development in the countryside, has been halted much sooner than expected, by economic forces rather than political initiatives.

6.4 Water-based recreation

The presence of water is often regarded as a fundamental requirement for outdoor recreation, either as a medium for the activity itself or to enhance the appeal of a recreational setting. (Pigram 1983, 133).

As far as the countryside is concerned, water is perhaps the single most important 'facility', after the land itself (Seeley 1973). The attraction of water as a backdrop has been a classical feature of the English landscape, as well as an integral part of most people's holidays and recreation outings, whether in terms of coast, lake, or river. Whilst naturally occurring, the creation or use of artificial water resources for recreation is also important, as witnessed by the after-use of mineral workings in the Thames and Trent Valleys or the creation of ornamental ponds and lakes in the grounds of estates and stately homes.

However, water is also one of the basic requirements for life. Often in the past thought of as 'free', or at least in bountiful and renewable supply, we have become all too aware of the fragility of our supply, as well as evidence of its declining quality. In addition, the power of water to flood and destroy life and property is never far from our consciousness. Indeed, the Thames Barrier, built to protect London from flooding has itself become a recreational attraction for East London. Equally, water is an important source of food, both in terms of salt- and fresh-water fish, as well as a host for a wide variety of fauna and flora. In considering the use of water for recreation, therefore, it is important to recognise that not only is it one of a variety of uses, but it is probably one of the least important, certainly in most tangible senses.

Four major categories of water-based recreation activities can be delineated:

boating, including sailing, rowing, canoeing, sail boarding, power boating, motor boating, cruising and water skiing;
traditional sports, including fresh- and salt-water fishing, sports fishing, coarse fishing and duck shooting;
swimming, including sub-aqua and surfing; and
land-based activities, including walking, rambling, nature study, jogging and picnicking.

From this categorisation it is possible to see the wide range of activities associated with water and the consequent range of people who might demand access to water. An alternative way of considering the potential uses of water for recreation is according to the resource itself. Patmore (1983) suggests a three-part classification, based on:

landscape, as a background, or for the land-based activities listed above;
node, including lakes and harbours for boating, swimming and some sports activities; and
corridor, such as streams and rivers for boating, swimming, some of the traditional sports and some land-based activities, such as walking and nature study.

In terms of demand, activities associated with water have been some of the fastest-growing in Britain, as well as some of the most popular. Angling and walking are two of the most common outdoor recreation activities in Britain, with a consequent strain being put on all resources that are close to large population centres. In contrast, some of the other activities, such as sailing, cruising and canoeing, are minority sports, but ones that have grown at a fast rate over the past 40 years. Indeed, club membership of the Royal Yachting Association trebled between 1950 and 1970, from 500 to 1500 club members, although it has remained relatively static since that time. Individual membership grew from 1400 to 30 000 from 1950 to 1970, doubled to 65 000 by 1980 and has increased by another 5 000 to 1991. Thus the pressure for suitable facilities has been increasing rapidly over the last four decades, and from some very small beginnings in the 1950s.

Against this background of considerable demand is the fact that Britain has few naturally occurring lakes and reservoirs outside of the Lake District and North Wales, as well as relatively few rivers that are navigable for significant parts of their length. This stock has been added to since the 1960s with a new range of reservoirs not only built to keep up with the ever-increasing demand for water (for agriculture and industry, as well as domestic supply), but also with a view to maximising their recreation potential. Simultaneously, Britain's canal network was revived to cater specifically for recreation, whilst road and property development led to the excavation of many mineral sites, primarily in the Thames and Trent Valleys, which eventually led to an increase in the stock of sites available for water-based recreation.

Quite apart from the water holdings themselves, water companies and river authorities traditionally bought large tracts of catchment land to surround their lakes and reservoirs. Often these were then let to the Forestry Commission for tree planting, but with strict controls over public access and pollution control. In some areas, like the Lake District, public access over water company land was allowed, if not encouraged, although in general public access was restricted.

These restrictions on access were mainly due to fears over pollution, since few catchment areas traditionally had any filtering or cleansing facilities. However, there has long been evidence that this fear was overstated. Even in reservoirs with primitive filters there is little pollution from humans, except where use gets very heavy over a small area of reservoir; in many cases modern filtering equipment can even cope with this level of pollution. As Hilson, in a publication of The Institution of Water Engineers and Scientists, points out:

> The use of reservoirs and associated gathering grounds for recreational purposes and for regulated access by the public should not significantly increase the pollution load arising from natural causes already carried by the reservoir (1981, 57).

Indeed, there is evidence that pollution from other sources is likely to cause more harm to recreation, through fish being killed or swimming being prevented, than vice versa.

The most serious problems caused by recreation concern flora and fauna, rather than just the quality of the water. Amongst these problems are the stocking, taking and breeding of fish, which affects not only the indigenous fish, but also water life and birds. Another type of problem is caused by the introduction of hydrocarbon oil products from the increasing numbers of boats on inland waterways, whilst in addition to the bacteriological effects of human contact with the water, increases in chemicals such as nitrogen and phosphorus could lead to increases in algal growth, necessitating removal by a clarification treatment process (Hilson 1981). Finally, effects such as trampling and erosion could occur on the banks of reservoirs and water courses due to increased public access. These and other environmental effects are discussed in greater detail in Chapter 8.

Regardless of evidence to the contrary, water companies and, latterly, water authorities, have been reluctant to allow recreational use of their water on the grounds of pollution control. Indeed, not until the findings of the Heneage Committee (1948) was sailing allowed on reservoirs, whilst access to reservoir banks, as well as bathing in the reservoirs remained prohibited. A slight relaxation of attitudes occurred through the 1950s, even extending beyond sailing to canoeing, but only on the understanding that the canoeists were proficient in capsize drill, in order to minimise their contact with the water! As Hilson suggests, bathing remains frowned upon, although not officially on pollution grounds:

... it is likely that swimming and related direct body contact activities would generally be banned, either on aesthetic grounds or in the interest of the safety of the bathers themselves (1981, 50).

In organisational terms, as Blunden and Curry (1989) note, the water industry has been subjected to regular upheavals since the end of the Second World War. The stated reason for this volatility was the rapidly changing needs of the population requiring a rapidly changing response from the industry. Closer to the truth, however, was the changing way in which successive governments wished to control and develop what was, after all, a vitally important national resource.

The most recent history dates from the Water Resources Act 1963, which created a three-tier system of water management, comprising 29 river authorities, 1393 sewage treatment departments of local authorities and 157 statutory water undertakings. Within this structure, the river authorities were responsible for the major navigations and catchment areas, the statutory water undertakers with delivering water to the consumers and the sewage departments with disposing of the results. Whilst keeping the functions separate, there was no effective national perspective. This led, understandably, to a somewhat parochial view of water supply.

Quite separate to these functions was the British Waterways Board, created under the Transport Act 1962. The British Waterways Board controlled, or was responsible for, about two-thirds of the public navigations in Britain, principally from the point of view of transport but also, since the Transport Act 1968, for recreation. However, since many of the navigations, and in particular the canals, were important for land drainage, the British Waterways Board inevitably overlapped with the river authorities and, to a lesser extent, with the statutory water undertakers.

In terms of recreation, the 1960s saw a further relaxation of attitudes, with more angling, sailing and canoeing taking place on reservoirs, although without any real encouragement from the water companies. The Water Resources Act 1963 gave the river authorities the power, if not the compulsion, to allow the public to undertake any form of recreation on their land. However, the river authorities invested little in recreation facilities and certainly did not plan for recreation at all, other than allowing use by some angling and sailing clubs. Eventually, in 1967, government pressure to increase the availability of water for recreation did bring about a classification of water resources. From that time water bodies were either of the direct-supply type, which were subject to severe restrictions, or they were for river regulation, in which case they were suitable for most recreation activities.

In contrast to the river authorities, the British Waterways Board never had a strong mandate for water supply and should have been, therefore, in a stronger position to develop the recreational use of canals and navigations. It responded to the Transport Act 1968 by setting up two divisions, responsible for freight and for amenity. It then divided its canals into one of three designations, according to the predominant use of the canal. The designations were commercial, cruising and remainder. As the names imply, canals were either for freight, recreation or they were redundant. The majority of its canals were classified as cruising waterways, to be developed and maintained for use by the public. Where appropriate, the British Waterways Board also attempted, by private Act of Parliament, to upgrade remainder waterways into cruising waterways in order to provide the most complete system possible. In some cases, such as the Kennet and Avon Canal, this is proving difficult due to the current financial status of the Board.

Regardless of these attempts to provide more adequately for recreation, the overall situation remained one of fragmentation and wasted opportunities. In an attempt to improve the situation the Sports Council decided that the preparation of regional plans for water recreation, designed specifically to make the most of the available resources, would be a suitable task for the recently formed Regional Councils for Sport and Recreation. However, no sooner had the majority of the Councils completed their plans (examples include East Midlands Council for Sport and Recreation 1972, Greater London and South East Council for Sport and Recreation 1971 and South Western Council for Sport and Recreation 1971) than central government, concerned over the administrative confusion caused by the 1963 legislation, reorganised the entire industry in the Water Act 1973. The 1973 Act replaced the river authorities, statutory undertakers and sewage departments with ten new unitary Water Authorities, nine in England and one in Wales. At the time this was considered to be a revolutionary move, in bringing the supply of clean water and the disposal of dirty water under the same management and control (Parker and Penning-Rowsell 1980). With regions based not on any existing regional boundaries, but on river basins, the water authorities represented what appeared at the time to be the first major step in implementing regional planning in England and Wales, the same already being true of the other British nations.

The new water authorities were given the responsibility for: the development of water resources; water distribution; pollution prevention; sewerage and sewage treatment; river management; flood protection; land drainage; sea defences; fisheries; and, latterly, recreation. Given a large degree of autonomy from local government, and particularly from the political framework of local democracy, the water authorities were really answerable only to central government, via the Secretary of State for the Environment. Initially, under a Labour regime, this led to the first attempts at national planning of water supply and distribution, with the strategic plans of the ten water authorities contributing to an overall national water plan, co-ordinated by the National Water Council, itself appointed by the Secretary of State for the Environment, under the Water Act 1973. In addition to considering the water supply and sewerage needs of the nation, the national water plan was also to include provision for the use of inland water for recreation and the enhancement and preservation of amenity. These latter functions came within the ambit of the newly formed Water Space Amenity Commission, charged with the duty of advising the Secretary of State for the Environment on the formulation, promotion and execution of the national water policy with respect to recreation and amenity. However, the Water Space Amenity Commission was given little money and, at its height, never had more than five staff, meaning that its impact was never likely to be great.

However, following the findings of a House of Lords Select Committee in 1973 (House of Lords 1973) that water recreation was unlikely to be self-financing, questions were raised over who was to pay for the recreation and amenity elements of the national plan. In the event this was never tested as a national plan was never produced. Whilst the water authorities were partly at fault for this, in not producing compatible policies, most of the blame lay with the ineffectiveness of the National Water Council and, as far as recreation was concerned, with the Water Space Amenity Commission (Greenfield 1975).

Consequently, few of the water authorities had a high priority for recreation and amenity. A notable exception to this was the Welsh Water Authority which produced a comprehensive strategy for water space, recreation and amenity (Welsh Water Authority 1980). In the main, how-ever, less than one per cent of authorities' budgets were devoted to recreation, and then it was mostly concerned with angling and fisheries protection.

In contrast to the water authorities, the British Waterways Board was spending at least half of its budget on recreation and amenity. However, most of this was simply in an attempt to keep canals open; even so, this was not enough to maintain, let alone improve, them. The 1970s and 1980s were, therefore, a poor time for inland water recreation. Not only was there ineffective provision and control, but subsidies of what provision there was from the general water rates meant that the authorities' recreation policies tended to be socially regressive. As Parker and Penning-Rowsell stated:

> There is no doubt that water recreation and amenity planning is the poor relation amongst the duties of the Water Authorities, leading to questioning of their suitability for the task (1980, 168).

More recent years have seen some improvements in recreation provision by the water authorities, especially with the construction of new multi-purpose reservoirs, such as Bewl Bridge and Kielder, where intensive recreational use can be accommodated on water supply reservoirs. However, severe disquiet remained about the general performance of the water authorities, particularly over their failure to produce plans and their failure to maintain adequate levels of investment in water supply and treatment facilities. Consequently it was to be expected that changes would be proposed. The most serious of these proposals was announced by the Department of the Environment in 1977, featuring a new statutory National Water Authority which would replace the British Waterways Board, the National Water

Council and the Water Space Amenity Commission and assume control for all planning and strategic management of water resources. However, the Conservative election victory of 1979 put an end to this proposal, as well as any notion of national water planning.

Following a further decade of mounting criticism about the water authorities, latterly concentrating on their financial performance and accountability, the Water Act 1989 ushered in a new era of water planning and delivery. In place of the water authorities came new privatised water companies, charged with water supply and waste disposal functions. Regulation and control of water supply and quality was, however, not vested in these companies, but in the newly formed National Rivers Authority. In this one piece of legislation the government created what has since been heralded as the strongest water control and conservation body in Europe, whilst at the same time subjecting water supply to market forces for the first time since 1963. This separation of prime polluters – the water companies – from policing and prosecution – the National Rivers Authority – has already led to a considerable tightening up of controls, as well as the prosecution of water companies for pollution offences.

Whilst the reorganisation of water supply and control functions appears to have been successful, there is less certainty about the future of recreation provision. Although the water companies have the responsibility to provide for recreation and access – and the National Rivers Authority is charged with ensuring that they discharge their responsibilities – pollution control is of much greater significance at present, especially to the National Rivers Authority. Of particular concern is the future of much of the land now owned by the water companies, a lot of which was traditionally used for recreation, as well as access to reservoirs and water courses. Faced with the dual needs of realising capital to invest in pollution control and showing the best financial return to shareholders, the water companies have been seeking to sell all excess land (with development potential where possible) as well as increasing charges for access to recreation activities.

These problems were, to some extent, foreseen as an inevitable consequence of privatisation. Indeed, in the committee stage of the Water Bill clauses were inserted in an attempt to protect peoples' rights of free access where they already existed, as well as placing covenants on any land that might be sold after privatisation. These provisions were of particular importance in the Lake District since private Acts of Parliament in the Nineteenth century had given rights of free access to water authority land; these were specifically protected in the Water Act 1989.

The inevitable result of the Water Act 1989 is that some former rights of access will be lost. Apart from the land in the Lake District National Park, as well as some land in other National Parks, there is really very little that can be done to protect former rights. Furthermore, even where elements of recreation remain, there is likely to be a change of emphasis away from informal activities with no financial benefit to the water companies, to new commercial ventures such as hotels and timeshare developments. Even the undoubted power of the National Rivers Authority is unlikely to protect former recreation usage. Quite apart from the rather more pressing problems of declining water quality and future water supplies, the National Rivers Authority is essentially a conservation body, meaning that recreation is likely to assume a very low priority in terms of staffing and, therefore, attention.

Contrary to many expectations, the British Waterways Board survived the Water Act 1989 as a public body presiding over an increasing number of usable canals and, reportedly, a growing number of canal users. The situation over canals is a complicated one, however. Whilst the British Waterways Board does not have the resources to renovate disused, or remainder, canals, many enthusiasts and preservation trusts, often with local authority and government training scheme finances, have succeeded in restoring whole canals, such as the Basingstoke Canal and the Kennet and Avon Canal. However, once restored considerable maintenance monies are required in order to keep the canals open, yet legally they remain classified as remainder waterways, so preventing the British Waterways Board from supplying any of the funds, even if it had them. In the short term this has been resolved for canals such as the Kennet and Avon by the local authorities agreeing to underwrite maintenance costs for the first five years, by which time the British Waterways Board should have been able to have them reclassified as cruising waterways.

In addition to the problems over maintenance and operation costs, there is also much fragmentation in licencing boats to use inland waterways. Since the British Waterways Board is not responsible for many river navigations, although canals generally connect such navigations, users can be faced with the expense and inconvenience of acquiring a range of licences for even quite short journeys. A particular case in point is the Kennet and Avon Canal, which connects with the Severn navigation at its western end and the River Thames at its eastern end. In order to negotiate the entire length, or to join up with other canals, users will require not only a British Waterways Board licence, but also one or both licences from Thames Water plc and Severn Trent Water plc.

An additional, although separate, problem associated with the increasing popularity of water recreation is the need for more associated facilities, such as car parks, slipways and marinas. Whilst traditionally thought of as coastal features, and still predominantly found in coastal locations such as the Solent in Hampshire, there is a growing demand for marinas on inland waterways. As Seeley (1992) points out, applications for marina development have met considerable resistance from local planning authorities, even in established areas such as the Solent, largely due to the scale of impact which they can have on the environment. Indeed, in addition to the area of water occupied by the marina, a similar amount of land will be necessary even for basic boat storage and repair, car parks and a chandlery, as at the

Beaulieu marina on the south coast or the Newbury Town boatyard on the Kennet and Avon Canal. Beyond this, local authorities have also been wary of marina proposals being used as a means of gaining more lucrative developments, such as retail and offices at Ocean Village, Southampton, industrial development at Universal Shipyards, Southampton, or housing at Hythe Marina Village on the Solent. There is as yet less evidence of marinas being used in the same way in inland locations, although they are often a feature of urban renewal programmes, such as the London Docklands development. Details of marina construction and development can be found in Seeley's (1992) book on public works engineering.

In concluding this subsection, there is an uncertain future for inland water recreation in Britain, brought about partly by the privatisation of the water companies and partly by the evident lack of interest in recreation by all but the recently renamed British Waterways. Ultimately, the future for much inland water recreation must lie in separating the responsibilities of the water companies from any non-commercial activities. Once this has been achieved they will be able to concentrate on commercial activities, including some elements of recreation provision. The remainder of the functions, including informal access and recreation, maintenance and licencing of the canals and rivers and the protection of amenity, should become the role of one statutory body, rather than being shared by a range of statutory and commercial interests as at present. If this were to become the case there is every reason to believe that recreation could attain a higher profile and that appropriate numbers and types of staff could be employed.

6.5 Conclusions

Claims by the Ministry of Agriculture, Fisheries and Food that 1988 was a highly significant year in terms of altering farming policy and the associated grant system (Ministry of Agriculture, Fisheries and Food 1989) cannot be denied. In addition to the Farm Diversification Grant Scheme, the Government also introduced new conservation and pollution grants, the Farm Woodland Scheme and the Set-aside Scheme. Although only the diversification grants were directly related to non-agricultural development, the latter two could certainly both be used as the basis for new, non-farming ventures.

In the first year of the new schemes, some £27m of private funds was invested in schemes qualifying for diversification grants, matched by £7.3m of government funds. In addition, some 650 farmers entered the Farm Woodlands Scheme, planting nearly 4600 hectares of trees and nearly 2000 farmers entered the Set-aside Scheme, taking almost 60 000 hectares out of production (Byrne and Ravenscroft 1989). At that time the relationship between land in Set-aside and land used for woodland or other non-agricultural activities was slight, with only ten per cent of Set-aside land being used for these purposes. However, it was recognised that the speed with which the new schemes were announced meant that few had time to prepare suitable plans. Indeed, it was estimated that up to a quarter of all Set-aside land was destined for a non-agricultural, often recreational, use in the near future (Byrne and Ravenscroft 1989).

Apart from the managerial and financial implications of the shift away from agricultural production, farmers also found themselves, often for the first time, falling within the ambit of the Town and Country Planning legislation. As Chick and Scrase state, in the introduction to their guide to the planning system:

> In the past farmers have been little involved with town planning and the planners. The use of land for agriculture does not need planning permission and the erection of a large range of farm buildings is also exempt. (1990, 3).

Equally, the planning system had never had much to do with farming, or the control of development on farms. As Chapter 5 and Section 6.2 indicated, even in the rural areas outside specifically designated areas there has always been a strong presumption against development. Thus there could have been considerable conflict between the countryside policies of the Department of the Environment and the Ministry of Agriculture, Fisheries and Food. However, it was apparent that the lack of emphasis previously placed on any form of proactive recreation policy meant that, or was a consequence of the fact that, recreation planning was the scene of administrative chaos. Six government ministries were involved, whilst most structure plans treated recreation as an afterthought, by allowing it where other land use interests did not dominate. Thus, recreation was not to be allowed in areas of high landscape beauty, nor on productive farmland, nor where people could cause damage. Furthermore, since structure planning is largely concerned with the development of facilities, and recreation is most often concerned with activities, such as walking, riding or cycling, there was very little commonality between government pressure to provide for recreation and the administrative machinery necessary to achieve it.

Alongside these policy shifts, both concerning the countryside and agriculture, the Countryside Commission decided that it should undertake a major policy review. This was called Recreation 2000 (Countryside Commission 1986). Whilst ostensibly concerned only with the recreation functions of the Commission, the review effectively challenged the previous supremacy given to conservation issues. Indeed, the outcome of the review, whilst being predictably narrow in not challenging the underlying issues of property rights, was startling in that the Countryside Commission moved almost immediately from being facilitators to being promoters (Blunden and Curry 1989). With its new policy document, 'Policies for Enjoying the Countryside' (Countryside Commission 1987), the Commission immediately called for a regional planning perspective to overcome the current chaos and produce a coherent system of recreation sites and facilities.

Even at the County level, often the scene of restraint in the past, there was considerable enthusiasm for promoting some development in the country-side:

> Whilst the farming community wishes to avoid too many developments which might detract from the County's environmental quality, it must be recognised that environmental quality owes a lot to the existence of a healthy farming industry. Consequently farmers must be permitted to develop their enterprises to sustain farm viability. (Hampshire County Council 1987, 9).

and:

> Marginal agricultural land and redundant buildings should be considered immediately for alternative enterprises. Limited residential development should not be ruled out and the employment opportunities afforded by allowing change of use should carry more weight than highway considerations or building regulations. (Hampshire County Council 1987, 10).

Whatever the official policy stance, or the relationship between national exhortation and local decisions, it became apparent that many new enterprises were gaining planning consents. The Council for the Protection of Rural England reported that in 1988 just four out of 960 applications under the Farm Diversification Grant Scheme were rejected owing to a refusal of planning consent (Anon 1989). Similarly, Ilbery (1987) has noted that a wide range of activities are gaining planning permissions in the countryside, although sometimes, as Hanson (1989) illustrates, with highly onerous conditions.

However, suggestions by government that many of these alternative uses should fall outside the planning system raised some vociferous objections, notably from the Council for the Protection of Rural England, which felt that such relaxation of control would allow the countryside to become a home for uncontrolled, noisy and intrusive developments:

> These could range from year-round clay pigeon shooting and war games to restaurants and shops, together with attendant car parks, toilets and the trappings of suburbia, all attracting large numbers of cars down unsuitable country lanes. The very suggestion that these activities might be outside effective planning control demonstrates an extraordinary insensitivity to public expectations of the planning system and the future of the countryside. (Anon 1989).

Faced with the beginning of what appeared to be a backlash against the plan to encourage a more diversified countryside, even the Countryside Commission was forced to temper its enthusiasm, by advising Ministers and others concerned with the future of the countryside that major developments should be strictly controlled, notwithstanding the need to generate new economic activity:

> Leisure and tourist developments can be in sympathy with their surroundings and indeed often depend on a beautiful and accessible countryside for their success. ... Schemes in the right place and of the right scale and design are welcome, but problems often arise because the development is inappropriate for the proposed site. Large holiday villages, marinas, theme parks and even golf courses have been proposed in sensitive areas of the countryside – imposed on the countryside rather

than being part of it. Major tourist schemes of this kind are in principle no different from other types of intrusive development in the countryside; the visual impact, traffic generation and other factors often correspond to that of a major industrial development. They are inappropriate in a sensitive countryside, especially designated areas. When allowed elsewhere, they should be subject to the highest standards of design and landscaping. (Countryside Commission 1989, 16).

In the face of mounting opposition, and amid a crisis of popularity, the Government reversed its attempt to relax the power of the planning system in the countryside (C. Patten, speech at the 1989 Conservative Party Conference). However, the extent to which this whole era of policy change and upheaval actually affected the nature of individual planning policies is open to some conjecture. Since the Countryside Commission's call for a regional planning perspective was never taken up, countryside planning was, and remains, largely the preserve of District Councils. By their very nature conservative and opposed to change, it is doubtful how many of them readily embraced the policy shift towards leisure development, even if they did grant some permissions in the period 1987 to 1989.

In the wake of the debate over countryside policies there is evidence of mounting interest and politicisation. The Church of England, in following up its controversial report on urban policy with a comprehensive review of countryside policy entitled 'Faith in the Countryside' stated that:

> In the 1990's... England's rural areas may be best understood as an arena – an arena in which different concerns and aspirations, stimulated by social and economic change, meet and must somehow be reconciled. (Archbishops' Commission on Rural Areas 1990, 4)

Equally, both the Sports Council (1990) and the English Tourist Board (1988) have found it necessary to develop strategies and policies for the countryside for the first time, whilst the Countryside Commission has recently published new consultation papers on the countryside (1991 and 1991A). The approaching unification of Europe has focussed attention once again on one such concern; the issue of free access to open country. In contrast to the position in the United Kingdom, many European states protect a constitutional right of public access to certain types of country. This has prompted the Labour Party to declare its support for a similar right in this country (Gould 1990), whilst a number of researchers are currently considering how such a right might be applied in this country.

In response to this the Government has so far gone no further than reasserting its support for the public rights of way network as the appropriate means of facilitating access to the countryside. As a result of earlier work on public rights of way, but introduced at an opportune time by Government, the Rights of Way Act 1990 seeks to ensure that footpaths and bridleways in arable areas are not obstructed by agricultural operations, as has often been the case in the past (Joint Centre for Land Development Studies 1985).

There is also evidence of an increased polarisation of attitudes to the countryside. In particular, there is increasing dissatisfaction with the operation of the National Trust. Although originally formed to protect common land for the urban poor, the National Trust has increasingly been annexed by the middle classes as a vehicle for preserving the social and physical landscape of Britain. Amidst accusations of being elitist and supportive of farmers and landowners rather than of the population as a whole, ill feelings were exacerbated when the governing council of the Trust decided to delay the implementation of a vote by the members at the 1990 general meeting to ban stag hunting from its land. By openly indicating its contempt for the rank and file, just as the landowning factions of Parliament have done for some time, the Trust has shown that the supposed progress made towards the democratisation of the countryside has been largely illusory.

The future of countryside recreation development is unclear. With the recent changes to the Nature Conservancy Council, now divided into national conservation bodies, together with a purported increase in environmental awareness on the part of government, at least as far as producing the recent White Paper 'This Common Inheritance' (Secretaries of State for the Environment *et al.* 1990), it seems likely that a greater emphasis will be put on conservation, possibly at the expense of recreation. The English Tourist Board, in its new strategy document (English Tourist Board 1991), confirms this view in making the point that, whereas over the last 20 years tourism and recreation have been 'principal players in national economic performance', such growth policies will no longer be acceptable in the future. However, it is equally clear that recreation, and associated spin-offs such as craft workshops and rural retail centres, provide one of the few realistic ways of promoting rural development and providing incomes to farmers and others previously reliant on agriculture for their livelihoods. Perhaps the greatest determinant, however, is the political persuasion of future governments and their attitudes to the primacy of freehold land tenure. Ultimately, regardless of policies to promote or regulate countryside recreation, it is the actions of private landowners that will determine the future nature of recreation development in the countryside.

7 Urban Leisure Provision and the Role of the Public Sector

7.1 Introduction

It is apparent from the preceding chapters that statutory planning has become largely concerned with the control of development. Whilst this role might be perfectly adequate for a market economy with no public provision or direction, it is clear that in a mixed economy there is a need to plan both market and non-market provision. This has inevitably meant that local authorities have, over the last century, occupied the joint roles of recreation provider and planning authority. Indeed, earlier chapters have shown how these roles have both developed from their common origin of concern for the physical health and moral condition of the working class (Glyptis 1989).

As the work on rural recreation in Chapter 6 indicated, the relationship between the ownership of resources and the provision of facilities is a very direct one. This suggests that, even if there had been stronger, more positive plans to guide leisure provision in urban areas, local planning authorities would still have been at the mercy of local authority leisure and recreation departments, since the latter effectively have the power to determine what facilities will be provided, and where (an example of this concerns the Guildford Leisure Centre proposal described in Chapter 9).

The potential effect of this dual role can be to put the planning committee and officers in an invidious position. This would be the case particularly where an authority decided that it required a new leisure centre on land that it already owned, regardless of the fact that a neighbouring authority had just built such a facility close by and the land proposed for the development was zoned for other purposes. Equally difficult, and possibly more divisive, is the case where the recreation department of the local authority has identified new facilities that it wishes to see developed to meet people's demands, but for which it does not possess the necessary capital. Rather than go without these facilities, it is in the interests of the recreation department to persuade the planning committee to allow major new private development in the area as long as the developers agree to provide the required leisure facilities. Whilst instances of such planning gain can often work well for both the community and the developer, the pressure on the planning committee can be immense.

A further point, concerning the direct provision of facilities by local authorities, is that the majority of rural recreation 'facilities', such as rights of way and areas covered by access agreements, do not require planning permission or any other involvement with statutory codes. Whilst it has become common to assume that this minimal requirement for facilities is a function of the types of leisure activity undertaken in the countryside, it is questionable whether urban areas are really any different. Indeed, regardless of area, the most popular out-of-home leisure pursuits, such as walking, jogging and fishing, do not require any formal or specialist facilities. However, the overwhelming response of government to meeting people's needs has been, on the one hand, a concentration on land use planning and, on the other, the provision of actual leisure facilities. Indeed, the range and scale of provision has been substantial, as was shown in Chapter 4. Although perhaps having reached a plateau at present, provision has consistently grown over the past three decades. In the period 1978 to 1981 the number of municipal sports halls grew from 400 to 600, swimming pools from 600 to 700 and golf courses from 150 to 200; in addition there were approximately 130 000 hectares of park and 49 000 hectares of playing fields (Torkildsen 1986). The current figures, given in Table 7.1, indicate something of a reduction from the 1981 levels, possibly due to the closure of a number of small facilities as they are superseded by larger ones, and by the emerging effects of compulsory competitive tendering (see below).

Notwithstanding the lack of facilities, whether public or private, prior to the onset of public building programmes in the late 1960s, it is questionable how relevant this response has been to the effective provision of leisure services. That is, how far has this type of provision by government been based on catering for people's need, as opposed to demand, for leisure services and, consequently, how relevant should the statutory planning process be to leisure provision?

In considering the role of the public sector as both planner and provider, therefore, this chapter will first examine its role as provider, particularly in relation to the social and economic justifications for provision, particularly in the aftermath of the Local Government Act 1988 and compulsory competitive tendering. The chapter will then proceed to consider the influence of the private sector and the ways in which local authorities can modify commercial provision and management.

7.2 The economic rationale for public leisure provision

In contrast to countryside recreation, government has long been involved in the provision and management of facilities to cater for urban sport and recreation. As Chapter 2 illustrated, throughout the Twentieth century this provision has gone largely unquestioned due to the various attributes attached to participation in sport and recreation. However, even the value of these attributes to successive governments did not prevent the Thatcher administration, in its review of the role of government in society, from

Table 7.1 Estimates of local authority facilities 1990–91

Facility	Metropolitan authorities (excl. London)	London	County councils	Non-metropolitan district councils	Total
Indoor swimming pools	91	36	2	219	348
Indoor leisure centres with swimming pools	55	23	15	281	374
Indoor leisure centres without swimming pools	91	22	38	287	438
Community centres	254	112	20	1 015	1 401
Golf courses	30	14	–	96	140
Parks and open spaces (hectares)	19 345	7 190	4 050	35 932	66 517
Theatres	16	18	9	120	163
Concert halls	15	4	–	61	80
Arts centres	2	5	10	25	32
Art galleries/museums	74	17	125	272	488
Allotments (hectares)	1 090	636	–	4 375	6 101

Note: 1990–91 is the last representative year before Compulsory Competitive Tendering changed the nature of data collection and, hence, numbers of facilities.
Source: Chartered Institute of Public Finance and Accountancy (1990).

questioning the efficacy of continued public leisure provision. Much of the present day government provision was established at the end of the Second World War, under the umbrella of the Welfare State. This corporatist approach to government was based on the concept of a partnership between commercial interests, the state and the community, such that economic growth could be encouraged and those responsible rewarded through individual profits whilst, at the same time, social stability could be maintained by the state (Thornley 1991). The three principal sectors of the Welfare State were, and remain, health and social security, housing and education. The basis for providing these services was unequivocally that of equity and social welfare, concepts since attacked as meaningless by the Thatcher administration. The services were established to rectify perceived market imperfections rather than as a competitive alternative to existing private suppliers. As such, their major orientation has been towards service provision, with little attention paid to quantifying or justifying their services in terms other than those upon which they were provided (a comprehensive discussion is contained in Organisation for Economic Co-operation and Development, Group on Urban Affairs 1986).

Public sector provision for urban sport and recreation was never formally part of the Welfare State, although it became thought of as such during the 1960s and early 1970s (Coalter *et al.* 1986) and did not suffer the same levels of reduced funding as were recently experienced in the other sectors of the Welfare State (Bramham and Henry 1985). Regardless of the status of sport and recreation provision, it shares many of the same origins and attributes as the other services and, consequently, the same major weakness of being unable to quantify or justify its provision in anything other than the service it provides. Indeed, it is probably rather more vulnerable to central government scrutiny, since sports-related agencies, such as the Sports Council, have failed to 'establish the political legitimacy of the welfare status of leisure provision' (Coalter *et al.* 1986, 159) and its reduction or cessation is hardly likely to be the subject of a major public outcry, especially if there is seen to be a trade-off between leisure provision and reductions in personal taxation.

It is here that the fallibility of public leisure provision becomes apparent. It has yet to be established that people have a need for leisure, let alone leisure facilities (Mercer 1973 and Dower *et al.* 1981), yet many such facilities (including sports halls, swimming pools, sports stadia, parks, playgrounds, rest gardens and allotments in urban areas) have been provided by local authorities over the last three decades (Seeley 1973), with considerable encouragement and funding from central government, via the Sports Council. However, when central government begins to question the provision of such facilities, whether on ideological or financial grounds, as has been the case since the 1979 election, it becomes apparent that whilst providers may claim to be clear as to their motives for provision, they have tended to neglect the justification of the provision of specific facilities, the consequent measurement of output and the quantification of the full cost of these facilities to the community at large, thus leaving such services vulnerable to reduction or cessation on the grounds that their value to society is, at the very least, questionable.

The recent advent of the compulsory competitive tendering requirements has, perhaps inadvertently, served to bring these deficiencies to the fore. Indeed, the interest shown by some commercial operators in running many of the large urban leisure centres tends to indicate that these facilities may be equally as appropriate to the private, as to the public, sectors (Thomas 1991). Furthermore, the lack of objective performance criteria for most of those facilities has given potential new operators a largely unrestrained remit upon which to base their tenders and, if successful, their future operating policy (Gratton and Taylor 1991). With this being the common, if not universal, case, the time has certainly come to re-examine the role of public sector sports and leisure provision, and the place that publicly managed facilities should have within it, for it is blatantly clear that many aspects of what we currently believe to be essentially public goods are being taken up with relish by the private sector.

In their work on the economics of sport and recreation, Gratton and Taylor (1985) attribute the level of state activity 'involved in subsidising and encouraging recreation participation' to a situation where the market mechanism is inadequate in allocating resources to the recreation market (Gratton and Taylor 1985, 10). Excepting the case of centralised, state-controlled economies, most authorities agree with this view, limiting the role of government to that of a residuary body ensuring that any market failures are alleviated by intervention and control:

> In a purely private market, the mix of available goods and services and the utilisation of resources is the expression of a vast number of economic decisions by households and firms. Government involvement can be justified when the market fails to provide the goods, services or other opportunities that society values highly. (Cwi 1982, 72)

There are two broad categories of market imperfection recognised by economists, named externalities and merit goods, although it might be argued that merit goods are a form of externality. An externality occurs when the utility of an individual is affected by a good or service and that effect is not explicitly accounted for by the market mechanism. In common with the main elements of the Welfare State, it has traditionally been presumed that leisure, recreation and sport produce largely positive externalities that go unaccounted for in determining their market price. Thus, if left to the market mechanism alone, a sub-optimal allocation of resources would occur, creating inefficiency in the working of the market.

Three principal forms of externality are associated with leisure provision. These concern health benefits, improvements to quality of life and option demands. Whilst recognising the lack of conclusive evidence on the external benefits to the nation's health bestowed by sport, Gratton and Taylor suggest that:

Sport provides both psychic and physical benefits to participants. Psychic benefits arise from the sense of well-being derived from being physically fit and healthy, the mental stimulation and satisfaction obtained from active recreation and the greater status achieved in peer groups. Physical benefits relate directly to the health relationship with active recreation. Exercise is a direct, positive input into the health production function. A person who engages in regular physical exercise is likely to live longer, have higher productivity over his working life, and have greater life satisfaction and improved quality of life than a non-participant, by being physically capable of carrying out more activities. (1985, 6)

Given that this is so, the state, as principal provider of health care, gains through reduced demands for its service, whilst employers can expect higher rates of productivity and less time away from work on the part of their labour force.

The second externality, concerning the quality of life, is an equally difficult one to support with proof, especially since it has yet to be proved that people actually have a 'need' for leisure, recreation or sport (Mercer 1973 and Dower *et al.* 1981). However, this has not stopped claims such as:

Simple facilities for recreation and sport for young people might help improve life for many who would otherwise be attracted to delinquency and vandalism. (Secretary of State for the Environment, 1977, 7).

Accepting the lack of evidence to support this view, Gratton and Taylor (1985, 10) suggest that this external benefit could be 'very large'.

The third type of externality occurs when people value the existence of something, have the option to use it, but do not presently do so. Under the competitive market mechanism, any benefit that is unpaid is also un-accounted-for. Thus, people may have an option demand which they value, but the facility in question, if faced with earning income, may close due to lack of use, so depriving people of their option.

The extent to which these claimed external benefits are not adequately reflected in the competitive market is, however, open to debate. In the case of the health benefits of sport the nation is, at most, a secondary beneficiary since individual, personal benefits must be considered the main utility to be derived from sport. The recent growth in private health care and fitness club membership tends to support this view. Equally, the psychic and physical benefits of sport noted by Gratton and Taylor (1985) are essentially personal, whilst even the public sector sells club membership as a way of internalising the effect of option demand.

Rather more persuasive as a genuine form of externality is the merit good since, although it is classed as an external benefit, it is one that is accrued by society rather than by individuals. Merit goods occur when the state believes that private preferences are distorted from a socially determined norm and, consequently, that social welfare can be increased by overriding an individual's view of the merits of a particular good. From the merit good perspective, there is no need to question the state of the private market or the distributional effects of government intervention since the decision is political rather than economic (Cwi 1982). Furthermore, isolated examples of merit goods are inappropriate since the social norms upon which supply decisions are based will differ between political ideologies. This inevitably leads to controversy when government intervention is made on merit good reasons, as Ng states:

> Given that the social objective should be a function of individual welfare, the only acceptable ground for recognising merit ... goods is the divergence of individual preference from individual welfare (1983, 286).

Under this strict definition the merit good argument would only exist to rectify ignorance and irrationality with, as Ng (1983) suggests, ignorance better rectified by education and irrationality left unaltered, as the prerogative of individuals. This, however, presupposes that there is complete political consensus and that what is best for every individual is necessarily best for the nation as a whole. Since neither of these conditions is satisfied in Britain at the present the merit good remains an important form of external benefit.

One further aspect of welfare economics that is significant in public leisure provision is the concept of public goods. A public good results when the external benefits created become so large that consumers cannot be prevented from obtaining them. A further aspect of the public good concept is that one person's consumption does not prevent another from consuming the product. If left to market allocation, therefore, the good or service will be underprovided since the demand will be characterised by the 'free rider' concept, where one consumer will wait until another person pays for the good or service, and will then reap the benefits. Thus, where an external benefit exists, or where a good or service is deemed to have merit, as is the case with leisure provision, a public good could be provided by government to overcome the market imperfections. An example of this would be a large park. The external benefits associated with health and exercise make the provision of the park desirable, regardless of the antipathy shown towards it by the market mechanism, yet the likely size of the park would make the exclusion of non-payers both difficult and expensive, while one person's use of the park is unlikely to infringe on another person's use, until congestion becomes apparent.

The result of these characteristics is that many goods and services in leisure, sport and recreation have aspects of public goods, but are neither purely public nor purely private. They also have associated externalities which, for the most part, have proved incapable of quantification, although techniques such as cost benefit analysis can reduce this significantly. Other means of reducing the number and scale of externalities have included the imposition of special pricing policies, such as increasing property taxes close to parks and open spaces, a source of much revenue in some US cities (Seeley 1973). Finally, publicly provided facilities have varying degrees of merit good attributed to them, depending upon the ideology of the government in power.

Originally, public leisure provision was centred on the external benefits of public health and hygiene. This was extended at the beginning of this Century to include collective consumption goods, such as parks and open space, due to their aspects of public goods such as non-excludability and non-rival consumption. International sporting success is, state Gratton and Taylor (1985), very close to being a pure public good which has the effect of encouraging people to participate in sport and recreation. The state has, therefore, always promoted sporting excellence, particularly in Olympic sports competitions and, latterly, by the provision of six centres of sporting excellence, administered by the Sports Council. Finally, the state has, through the Sport For All programme (see Chapter 4.2), promoted the provision of indoor leisure facilities which, although essentially private goods themselves, have enabled government to promote participation in sport and recreation and raise the consumption of these private goods above the market-determined level.

The underlying feature to emerge from these aspects of market imperfection is the characteristic of intangibility; the reason why the market appears to be inefficient in the first place. However, this same intangibility that contributed to the market imperfection can also threaten continued government intervention; as Mulcahy remarks, with respect to the arts:

> ... public arts policies need a stronger basis for existing than vague, often unsupported, assertions that art is an incontestable public good. (1982, 33).

Furthermore, the complexity of the motives for government intervention on these grounds leads to a wide range of provision, from heavily subsidised community facilities such as the standard municipal park or swimming pool, to facilities that are publicly provided or encouraged, but financed privately, such as some new squash and tennis centres (like the Royal County of Berkshire Racquets and Health Club in Bracknell) and leisure pools (such as Rivermead in Reading). Even within government-provided and subsidised facilities there can be elements of the market, such as commercially operated retail and catering outlets within local authority sports facilities, such as the Oasis Centre, Swindon and the Coral Reef in Bracknell. Equally, the government can encourage some non-commercial provision within an otherwise private facility, such as the Countryside Commission and local authorities grant-aiding the provision of education, information and warden services at privately-owned recreation enterprises such as the Weald and Downland Open Air Museum near Chichester. Many local authorities also 'internalise' the option demand externality through the sale of facility or service memberships, such as the Reading Borough Council Passport to Leisure scheme, whereby local residents can purchase the right to future consumption at reduced fees, just as they might in the commercial sector.

This leads to the conclusion that the primary determinant of public policy towards sports and leisure provision and, consequently, towards the response generated appears to be the ideology of the government in power, and its consequent view of welfare considerations and the market mechanism. At one end of the spectrum, a centralised collectivist economy would expect all provision to be planned and carried out by the state, whereas the opposite would be expected in a free market economy. Indeed, the reduction in the size and scope of the public sector in Britain that has been a characteristic of the Conservative government elected in 1979 has stemmed from an ideological belief in the allocative efficiency of the free market, rather than a wish simply to reduce costs and increase efficiency (Walker 1984). This view is corroborated by LeGrand and Robinson (1984), who feel that the public sector in Britain has come under attack, as much for providing goods and services that are supplied privately in some other western nations, as for any inefficiency that may or may not have existed. Indeed, in his case for public support of the arts, Mulcahy states:

> What can we say to sum up the case for public culture? The most basic argument rests on the belief that culture is good-in-itself: what economists would call a 'merit-good' and what political scientists would call a 'value'. (1982, 54).

Thus, in both determining the nature of public leisure provision and the ways in which this affects the type of facilities provided, it is necessary to examine the aims established for leisure provision by recent governments in Great Britain.

7.3 The allocative efficiency of public leisure policy

In order to determine how far the Sports Council was right to have pursued this facility-orientated policy, it is necessary to determine what type of facilities can be directly associated with the success of the Sport For All policy. Since the Sport For All policy is so unequivocally based on people and their values, motivations and needs, an appropriate starting point might be to consider how far sport and recreation can contribute to people's lives, to what extent physical facilities are necessary to allow or encourage people to participate and, in turn, how far the facilities provided by the state are appropriate.

One method of analysing peoples needs and the ways in which they can be met is Maslow's hierarchy of needs (Maslow 1970). The basis of Maslow's hierarchy is that each individual has five basic needs that motivate his or her actions and responses to life. It follows, therefore, that for sport and recreation to play an integral part in satisfying people's needs, it should feature in the hierarchy. Equally, the Sport For All policy should equate with the hierarchy if it is to be appropriate in enabling people to satisfy their needs. Finally, if these conditions are met, it should be possible to establish the type of facilities that are appropriate to meeting people's needs and, therefore, the extent to which the state has correctly identified and implemented its policy for leisure provision. The relationship between Maslow's hierarchy and the Sport For All policy is outlined in Table 7.2.

Table 7.2 The relationship between Maslow's hierarchy and the sport for all policy

Maslow's hierarchy	Sport for all aims	Political/economic justification	Method of implementation	Personal motivation
1. Physiological need for food and water	Physiological need for food health and fitness	A belief in the intrinsic worth of individual health and fitness to the nation as a whole. Also explicit support of the relationship between improved health and reduced coronary heart disease	Provision of public sports and leisure facilities. Statutory provisions for physical education in schools. Marketing campaigns such as Sport For All	Basic consumption of food and health care for survival. Growth of private health and fitness clubs indicates the positive utility of personal consumption
2. Safety	Quality of life	Belief that healthy citizens have a more positive view of their life and their place in society	Legitimation of sports and leisure activities through schools and campaigns such as Sport For All	Positive association of sport with health, and health with quality of life
3. Affiliation	Social benefits of sport	Attempt to ameliorate the boredom and subsequent delinquency of some members of society; to provide an affirmation of their citizenship	Provision of, and support for, neighbourhood sports facilities, sports motivators and community leaders	Consumption utility of participation
4. Esteem	Psychological benefits	Reinforcement of 'quality of life' justification	Provision of multi-purpose leisure facilities, sports coaches and centres of excellence	Consumption of participation combined with personal utility of achievement
5. Self actualisation	Increased participation	Externality effect of encouraging others to emulate national sporting success	Provision of a variety of sports facilities, provision, training and support of coaches, and support for national and international competition	Personal utility of reaching or maximising potential

The first level in the hierarchy is the physiological need for food and water. Until satisfied, this need dominates human motivation and precludes the activation of the remaining need levels. Similarly, in the Sport For All policy, it is assumed that the physiological need for health and fitness is dominant until satisfied, since further sports activity is prevented by ill health, injury or lack of fitness. Gratton and Tice state that:

> Sport is different from other leisure activities in that it is supposed to contribute to health. There is a growing body of epidemiological evidence that sport and exercise do have a significant positive effect on health status (1987, 26).

This is confirmed by Fentem *et al.* (1979) who list over 1300 references on the relationship between exercise and health, by Paffenberger *et al.* (1986), who found that the death rates of active people are significantly lower than for non-active people, and by Fentem and Bassey (1978) in their report to the Sports Council, who claimed that exercise is good for everyone.

After physiological needs comes the second level, termed safety needs by Maslow (1970). These needs include security, stability and protection. Although there is no direct equivalent in the Sport For All policy, the quality of life aim is similar, in that after health, we require knowledge of the quality, stability and security of our existence. Riddick and Daniel found that:

> Leisure activity participation emerged as the strongest contributing factor to the life satisfaction of older women. (1984, 146).

In particular, life satisfaction increased as participation in leisure activity increased, a finding echoed by Russell (1987), who found that the quality of life of retirees is increased by access to, and participation in, a wide range of leisure activities.

The third level of need is termed affiliation needs. At this level we require love, affection and belongingness. Similarly, at this level in the Sport For All policy is the range of social benefits, centred on the affirmation of citizenship through sport. The use of leisure provision and the encouragement of leisure activity as an affirmation of citizenship is an important aspect of state leisure policy (*Coalter et al.*, 1986, Scarman 1981, and Department of the Environment 1977), and has led to the provision of leisure facilities and an expansion of leisure opportunities in areas of unrest or social disharmony (such as Brixton in London and Toxteth in Liverpool, described by Adie 1985).

Having satisfied the third level of need, the fourth level is activated, termed esteem needs. Within this categorisation would come feelings of self-confidence, worth, strength and capability (Mills 1985). The equivalent aim in the Sport For All policy is to produce the range of psychological benefits such as the enjoyment of leisure and the advancement of personality. Tinsley and Johnson (1984), in their taxonomy of leisure activities, state that psychological benefits are a strong motive for undertaking physical exercise, and are associated with 'expression through activity'.

The fifth, and highest, level of need is termed self actualisation, and is defined by Mills as 'the desire to become what one has the potential of becoming' (1985, 185). The corresponding aim of the Sport For All policy is that of increasing participation and improving performance in sport, since both these aims encourage individuals to seek their potential in sport, either by taking up a new activity, or by having the facility to excel at their chosen sport.

The three basic aims of the Sport For All policy are, therefore, concerned with the social values of health, security and citizenship. The fourth and fifth levels are, by contrast, essentially private values, based on the fulfilment of personal ambitions and fuelled by personal goals and motivation. They are, thus, largely voluntary in their social context, since they may contribute to the health, security and citizenship of that individual, but only as an externality associated with the activation of the fourth and fifth levels of need.

7.4 The role of leisure facilities in the fulfilment of the Sport For All policy

In considering the role of leisure facilities in fulfilling the Sport For All policy it is necessary to assess the extent to which they are required in order to achieve the aims of the policy. At the lowest level, the improvement of the health of the nation, there is general agreement over the link between sport and health. However, the optimum amount and kind of exercise for each individual remains ill-defined, although medical evidence suggests that it would centre on simple activities, such as running, cycling and swimming (Fentem and Bassey 1978). Provision for these activities could, therefore, have value in fulfilling the Sport For All policy, but only in as far as providing basic running and cycle tracks and swimming pools in sufficient numbers to allow all members of society to participate in these activities.

In considering the impact of the Sport For All policy on the quality of people's lives, McIntosh and Charlton comment that:

> Sports have improved the quality of life of some people in their own view ... but Sport For All has not been generally recognised as a necessary or a sufficient condition for a high quality of life. (1985, 11).

It would appear, therefore, that the provision of a range of leisure facilities to allow participation in a wide range of activities may increase the quality of life for some people, but that the value of this provision to society is unclear, since leisure activity is not considered to be a prerequisite condition for a high quality of life.

In analysing the place of leisure in the need for citizenship, Allen and Beattie (1984) found that whilst leisure activity proved to be a good predictor or signal of overall community satisfaction, it did not rank highly as an important aspect of satisfaction with community life. Indeed, the participants in Allen and Beattie's (1984) survey ranked leisure provision and activity below economic security, education, health and safety and the environment. Similarly, the Sport For All policy aim of having sports provision treated as a social service has not happened administratively

(McIntosh and Charlton 1985), since the provision of sport by local authorities remains non-mandatory and is financed by non-key sector funding. However, they did find that spending on leisure provision by local authorities has risen more rapidly than on housing, health or education over the past twenty five years and suggest, therefore, that it could be argued that this aim has been partially achieved. Whether this inference can be drawn is debatable, since the relative change in spending could be for a variety of reasons wholly unrelated to the aim of sport becoming a social service.

In considering the lowest three levels of Maslow's hierarchy of needs, therefore, the provision of leisure facilities does feature as an element of welfare policy, but only for basic activities associated with health and exercise, rather than any wider definition of sport or leisure. Thus, at these levels, the provision of basic leisure facilities has a potential value to society in connection with the improvement of health, security and community.

Whilst it is undeniable that sports participants motivated by esteem and self-actualisation, the fourth and fifth levels of the needs hierarchy, are likely to have already gained the basic benefits of health, security and community, and may be stronger members of society through their sports participation, it is questionable how far the provision of facilities to meet these higher motivations contributes to social welfare. This is an especially critical question since such facilities are likely to be the more complex, expensive modern leisure centres rather than the basic parks, running and cycle tracks and swimming pools associated with the lower orders of motivation. Indeed, it would appear that in a market economy, the level four and five motivations feature essentially private goods with little welfare externality attached to them. If this is the case, access to participate in such activities can be, and already is, purchased in the market place, rather than being provided by the community.

Thus, there is a strong argument that people's leisure needs, as classified by Maslow, are not, in the main, met by the majority of local authority-provided facilities. If this is the case then the traditional welfare economics arguments for the public provision of such leisure facilities are not strong. This is not to deny the external benefits associated with sport, as defined in the lower order levels of need, but to recognise that many people would, and already do, acquire these in the market in order to satisfy their higher, essentially private, orders of need. The principal justification for state intervention in the leisure market rests, therefore, on the externality and merit good arguments associated with the provision of basic facilities and encouragement for those who do not currently participate, or who do not want to, or cannot, progress beyond the most rudimentary levels of fitness.

7.5 The distributional equity of state intervention in the leisure market

Having already questioned the allocative efficiency of the full range of state leisure provision, there may still be a strong equity justification for provision, based on the types of facility provided and the types of people who benefit from them. The question of the type and number of facilities in Britain was one of the principal motivations for first establishing the Sports Council. In its document 'Provision for Sport' (Sports Council 1972), the Sports Council noted the deficiency of facilities and set targets for provision, to be achieved by 1981. All facilities were to be of the basic variety, with no attempt to provide for high level sport. A considerable amount of development took place over the next decade, mostly by local authorities grant-aided by the Sports Council. However, a shortfall of sports centres still existed in 1981, leading the Sports Council to concentrate on them in the following period.

In order to hasten the establishment of new facilities, the Sports Council took the initiative in encouraging the refurbishment of old buildings, the community use of school sports facilities and the production and promotion of a small standardised sports hall. However, whilst the number of facilities continued to grow, they also tended to become the more sophisticated 'leisure' centres, featuring retail and catering outlets in addition to sports facilities. Equally, new forms of swimming pools, including wave machines, flumes and fun pools were also developed in the public sector. In all cases the new facilities were justified on the grounds of encouraging more people to participate. They also represented a way of increasing revenue through increased fees and visitor spending, with 'commercial' management replacing the more traditional role of the public sector. Rather than abandon its role, however, the Sports Council continued to support this increase in facilities, with grant aid as well as technical advice and information. Thus, even in the mid-1980s nearly half of the Sports Council's budget was being devoted to facility provision and it was not until very recently that a change of emphasis away from facilities was indicated.

Although one of the supposed cornerstones of the Sport For All policy was increased and diversified participation, no records of participation trends were kept until the mid-1970s. However, since then it has been found that sport and active recreation have remained minority activities, even amongst the most active sectors of society. Thus even walking, the most popular activity, undertaken on summer evenings and weekends, the most popular times, still only regularly involves 20 per cent of the population. This has led McIntosh and Charlton to state that:

> The general conclusion must be that 'Sport For All' has done little to overcome the inequalities already present at the start of the campaign. Men are still considerably more active than women ..., participation decreases with age and with lack of education. Those with a car participate more than do those without and professional and managerial workers take part more often than do manual workers (1985, 82).

Indeed, since participation by some groups, such as the unemployed, actually declined after the advent of the Sport For All policy (McIntosh and Charlton 1985), it could be argued that the inequalities of market allocation have been exacerbated by state intervention. However, it is certainly clear that the bias towards facility provision has done little to ease access to sport for those who were, and remain, under-represented in sports participation. It has also allowed those who would have been active in any event to participate more, and at a considerably lower cost than they might otherwise have been prepared to pay. As Gratton and Taylor state:

> Even if we regard recreation as a 'need', it is naive to expect that building facilities, and offering recreation services at subsidized prices is an ideal way to cater to that need. ... the inefficient way to spend public money is to subsidize every consumer, since some consumers will consume above the desired minimum, without subsidy. It is not only inefficient but also inequitable when the main beneficiaries of the subsidies are the better-off. (1985, 215–6).

The most recent history of urban sports provision has seen the Sports Council still seemingly unaware of its failings. This was particularly graphically illustrated when it embarked upon establishing its corporate identity just as the whole of public sector leisure provision was being put in jeopardy by a government committed to reducing the level of public spending. In managing to display how out of touch it had become, the Sports Council still believed in 1985 that it and the Regional Councils for Sport and Recreation wielded some power when it stated, as part of its first corporate plan, that:

> Regional Councils for Sport and Recreation will identify and meet the most important needs of their regions. (Sports Council 1985).

How the Regional Councils were to play an executive role, or where they were to get their funding from was never made clear.

Even whilst the Sports Council were still debating how to provide yet more facilities, the Audit Commission had earmarked public recreation for an efficiency audit. Not surprisingly the outcome of this audit was that there was plenty of scope for increasing efficiency; and that this could best be achieved by subjecting all local authority leisure provision to compulsory competitive tendering. The ignominy of the Sports Council's decline has recently been completed by parliamentary debates over the future funding of recreation provision by a national lottery and by the removal of the Sports Council from the Department of the Environment, returning it to the control of the Department of Education and Science, where it started out as an advisory body some 25 years previously.

The evidence from a growing body of literature on the value of leisure activity to the welfare of society is that the provision of basic facilities for running, cycling and swimming, when associated with motivation to participate and support for participants, (through, for example, coaching, creche facilities and subsidised entrance fees), can lead to significant, if as yet largely unmeasurable, benefits to society. These are principally in terms of improved health, quality of life and community affiliation. What is far more contentious, however, is the value of providing large scale, complex leisure facilities. Even claims that modern multi-purpose leisure facilities may promote increased participation through providing for a wider range of activities are dubious, since participation remains steadfastly concentrated on certain user characteristics and any increase in participation witnessed since the advent of the Sport For All policy has been largely as a result of the current users participating more, rather than as a result of new participants being attracted to sport. Furthermore, even if the benefit of any leisure activity is improved health and fitness, it is questionable whether large expensive facilities can achieve more than smaller, cheaper ones, given that the emphasis of the cheaper facility is likely to be more closely associated with health and exercise.

The difficulty in evaluating the performance of existing facilities, characterised by the lack of operational aims and objectives may, therefore, be rather more appropriately characterised as a failure of existing performance measures to signal that certain leisure facilities are not appropriate to public provision, in spite of their, largely spurious, welfare claims. Indeed, the speed and relish with which many such facilities have been tendered for and, in some cases, taken over by private commercial operators, complete with newly-defined, market-orientated objectives, provides ample support for this proposition (Lynch 1989).

It is not so much that the policies enshrined in the Sport For All campaign have dubious legitimacy, therefore, but that they have not been implemented in the most appropriate manner. On the surface, this appears to be particularly so of the wide range of facilities which the public sector has provided and the ways in which participation has been encouraged. These two elements are, however, linked. Whilst the public sector provides what are essentially private goods at subsidised prices, the demand from those who would willingly pay more to participate is such that forms of rationing, via pricing, timing or club membership, mean that other people are effectively excluded. Thus until the type of facilities provided and the means of restricting access to previous non- or low-participants are clarified the beneficial elements of the Sport For All policy are unlikely to be experienced by a wider section of the population.

7.6 The compulsory competitive tendering of local authority leisure facilities

Following the Thatcher administration's determination to reduce the role and power of local government, culminating in the Local Government Act 1988 and compulsion to offer most services to the private sector on a competitive tender basis, all local authorities have been forced to examine what they do, or think they should do. In the past, note Gratton and Taylor (1991), few authorities or managers have made their aims or agendas explicit, implying little or no performance monitoring or evaluation. When faced with the need to write a specification for a service that is to be offered for tender, however, those same authorities and managers require objective measures against which performances can be evaluated.

Whilst the establishment of performance criteria has not proved too problematic for many technical services such as refuse collection, the definition, let alone measurement, of the social outputs of leisure services has presented local authorities with problems (Bovaird 1991). At present most authorities have been able to do little more than rely on explicit input measures, with only the most forward-thinking authorities attempting to specify intermediate output measures such as target participation rates. The upshot of this, in most cases, is that in the absence of output measures local authorities will be forced to rely on economy as the major determinant of the tender process, regardless of how poor, or even counter-productive, this is likely to be.

Regardless of these shortcomings, it is now inevitable that competitive contracting will occur for most leisure services. Even the political demise of Mrs Thatcher will not halt this process, especially when the most contentious current political issues surround health care and local taxation. What is becoming clear, however, is that in many ways Britain is already moving into a post-Thatcher era. Whilst the rhetoric of the new Prime Minister hides sentiments that are probably little different from Mrs Thatcher's, it is clear that her policies are deeply unpopular with a majority of the population. Assuming that this unpopularity will lead to a change of emphasis in government, or even a different party in power, local authorities have a chance to redeem the situation into which they have unwittingly allowed themselves to fall. Although there is little prospect of influencing the majority of leisure management contracts currently on offer, most are for a relatively limited period of time, whilst there is evidence that not all of the facilities offered will find new managers and of those that do, the vast majority will be operated by former council employees now working in direct service organisations (Thomas 1991 and 1991A).

However, relieved of the burden of necessarily operating such facilities, local authorities should now be in a position to re-define their own objectives and decide what facilities they should provide to complement those now catering predominantly for the private market. Indeed, as more people continue to acquire more leisure time, the basic facilities required to allow the nation to remain, or get, fit are becoming ever more important. Much work remains to be done on establishing the links between active recreation, facilities and the primary motivational needs of society. In the meantime, however, government can be sure that the provision of simple, low-cost facilities for swimming, running and physical exercise, when managed with straightforward, welfare-orientated policies, are likely to contribute significantly to the health production/consumption function and the welfare needs of society.

Faced with the realities of a market, albeit rather contrived, for recreation provision, local authorities will be presented with some new challenges, both in terms of their role as recreation provider and planning authority. In particular, recreation and leisure departments face the challenge of a redefined range of responsibilities including the letting and monitoring of contracts; duties allied to the ownership of property, such as ensuring that adequate maintenance regimes are enforced; some direct management if no suitable outside contractor can be found; and, most importantly of all, addressing the needs of the community to ensure that a full range of facilities is available, whether provided by the public, private or voluntary sectors of the economy. As a planning authority, it is likely to be faced with increasing internal pressure from the recreation and leisure departments to safeguard the place of recreation facilities in the local economy, whilst recent evidence suggests that external pressure is growing from developers keen to build new large-scale leisure developments in both urban and rural areas. Perhaps the most significant question to ask is whether this change in the role of local government, accompanied by the necessary shift in responsibilities, will allow it to discharge its duties more effectively in the next decade. As Parkinson notes, with respect to urban strategies in general:

> Is investment in physical, as opposed to human, capital the best way of ensuring urban regeneration? Can the market and the private sector address issues of equity and fairness as well as those of efficiency and wealth creation? What will be the consequence of the growing centralisation of government power ... and the loss of direct local accountability? (1989, 439).

The meagre evidence available so far indicates that the situation is far from lost; that a number of local authorities have determined ways in which they hope to protect equity and fairness whilst giving commercial operators the freedom of day to day managerial responsibility. The degree to which they are successful is, however, critical, for the great strides achieved in widening access to recreation facilities and recognising the part that recreation can play in people's lives could all-too-easily be lost in the commercialisation of basic recreation provision.

8 The Environmental Assessment of Recreation Development

8.1 Introduction

Ecology is like universal suffrage or the 40-hour week: at first, the ruling elite and the guardians of social order regard it as subversive, and proclaim that it will lead to the triumph of anarchy and irrationality. Then, when factual evidence and popular pressure can no longer be denied, the establishment suddenly gives way – what was unthinkable yesterday becomes taken for granted today, and fundamentally nothing changes. ... Ecological thinking still has many opponents in the board rooms, but it already has enough converts in the ruling elite to ensure its eventual acceptance by the major institutions of capitalism. (Gorz 1987, 3).

Whilst it is now some 30 years since the Western world was first alerted to the potential environmental impacts of economic development, by Rachel Carson in her book *Silent Spring* (Carson 1962), most people would be forgiven for thinking that environmental concern was a product of the late 1980s. In Britain this has largely been the case, with the Government heralding its White Paper on the environment (Secretaries of State for Environment *et al.* 1990) as the first comprehensive strategy of all aspects of environmental concern. Similarly, both the Labour and Liberal Democrat parties recently published policies on the environment in the expectation that it will be an important issue in the forthcoming general election (Labour party 1990 and Liberal Democrats 1990).

However, whilst the environment may have only arrived on the political agenda in Britain in the late 1980s, it has become much more widely accepted in many other countries, both in Europe and elsewhere. Whilst there was no mention of the environment in the Treaty of Rome, the Council of Europe did establish a Committee for the Conservation of Nature and Natural Environments in 1962 (Haigh 1987). Quite what effect this committee had is unclear, for it certainly did not have any executive power to implement conservation measures. In the USA the rising political salience of the environmental lobby in the 1960s, together with the recognition that standard methods of investment appraisal, such as cost-benefit analysis, had tended to devalue the environmental costs and benefits of government projects (Rees 1990), led the United States Government to pass the National Environmental Protection Act in 1969. Whilst only applying to Federal Government projects, the Act did assert the Government's duty to safeguard the environment for present and future generations. Initially, the way in which environmental assessments were carried out varied from State to State, to the extent that projects that were rejected on environmental grounds in one State were often allowed in a neighbouring State. This led to the formation of the US Council for Environmental Quality which developed an acceptable standardised form of environmental assessment for use in all States.

Environmental awareness progressed more slowly in Europe, as well as being perceptually limited to the notion of conservation. Thus, in the year following the enactment of the National Environmental Protection Act in the USA, Europe celebrated 1970 as European Conservation Year. This was followed, more significantly, by the United Nations Conference on the Human Environment, held in 1972 in Stockholm. Out of this conference came some agreed international environmental policies that were taken up with relish by the European Community. In the first of a series of action programmes agreed by the European Council of Ministers in 1973, four areas of work were identified:

1) reduction of pollution and nuisances;
2) non-damaging use and rational management of land, the environment and natural resources;
3) general action to protect and improve the environment; and
4) action at the international level.

One of the first specific actions linking the impact of recreation and tourism on the environment was initiated in 1977 by the Organisation for Economic Co-operation and Development with the formation of its Group of Experts on Environment and Tourism. After two years of meetings and many commissioned case studies, the Group produced its final report in 1979, recommending that environmental considerations be fully integrated into the tourism development policies of member countries at the earliest possible time (Organisation for Economic Co-operation and Development 1980).

Further action programmes from the European Council of Ministers followed in 1977 and 1982, although their effectiveness in achieving environmental protection and improvement was inevitably compromised by the Council's lack of executive powers, a problem shared by the OECD. As Haigh notes, with respect to the action programmes:

> The action programmes are very comprehensive but the instruments for putting them into effect are limited when compared to the means available to a nation state. The European Community can enunciate principles; it can commission studies and research; it can make grants from a small experimental fund for environmental purposes; it can set up advisory committees and otherwise exhort Member States and people within them by promoting meetings, conferences and publications; above all it can legislate. What it cannot do itself is directly to administer its environmental policy since it has no staff in the Member States. (Haigh 1987, 379).

Increasingly, the basis of these action programmes was an attempt to halt pollution and environmental degradation at source. This effectively meant that the European Community became committed to modifying development control in order to ensure that environmental impact assessment took place prior to the construction or development of any new project (Fortlage 1990). When the Member States did not follow the thinking of the Council of Ministers on this point, it was inevitable that the Council would eventually use its legislative power to bring them into line. This happened in 1985, in EEC Directive 85/337, whereby all Member States were obliged to integrate formal environmental assessment procedures into their existing development control legislation.

In Britain, Directive 85/337 was introduced into the legislation through the Town and Country Planning (Assessment of Environmental Effects) Regulations 1988 (SI 1988/1199). This meant that the submission of an Environmental Statement became part of a normal application for planning consent for a range of specified developments covered by the Town and Country Planning legislation. However, in bringing environmental assessment into the British legislation in this way there was no opportunity for the issue to be debated in Parliament, with the inevitable consequence that the Council of Ministers' preoccupation with pollution control was transferred to the British development control system, perhaps to the detriment of being able to consider the effect of development on employment, social structures and local economies (Fortlage 1990).

8.2 Environmental assessment in practice

The basis for determining whether or not an environmental statement will be a necessary prerequisite for any particular development is contained in Schedules 1 and 2 of SI 1988/1199. Planning applications for projects listed in Schedule 1 must be accompanied by an environmental statement, whilst applications for those in Schedule 2 need only be accompanied by an environmental statement if they are likely to have significant effects on the environment due to their size, nature or location. In marginal cases a developer can apply to the relevant planning authority for a determination as to whether or not an environmental statement is required.

Projects listed in Schedule 1 include major refineries, power stations, mining operations, ports and waterways (a full list can be found in SI 1988/1199, Department of the Environment and Welsh Office 1989 and Fortlage 1990). Schedule 1 is, therefore, solely for projects that are so large, obtrusive or potentially dangerous that it is imperative to have the fullest information about their likely environmental impact before development at a particular location begins. Since even the largest recreation and leisure projects are unlikely to be this environmentally significant, there are no such projects listed in Schedule 1.

Schedule 2, on the other hand, is a much broader collection of projects, including industrial estates, urban developments, ski or cable lifts, roads, harbours and aerodromes below Schedule 1 standards, dams, canals and flood relief works, surface or underground passenger train or railway services below Schedule 1 standards, pipelines, long distance aqueducts, yacht harbours, holiday complexes, motor racing circuits and any other noisy or obnoxious projects (taken from EEC Directive 85/337). In the UK these items are further refined by size and location standards. Thus, for example, a proposed industrial estate will be subject to a possible environmental statement if it is over 20 hectares in size, if there are more than 1 000 dwellings within 200 metres of the proposed boundary, or if it is in a sensitive location, such as an Area of Outstanding Natural Beauty. There are no specific guidelines for recreation developments in the UK, although any project covering more than 100 hectares falls within Schedule 2.

In terms of the actual process of environmental assessment, the Department of the Environment and Welsh Office, in their explanatory guide describe it thus:

> The term "environmental assessment" describes a technique and a process by which information about the environmental effects of a project is collected, both by the developer and from other sources, and taken into account by the planning authority in forming their judgement on whether the development should go ahead. (1989, 3).

Environmental assessment is, therefore, based on the systematic collection, analysis and presentation of information that should be available to the developer of any project. Whilst this may initially appear to be placing an undue burden on the developer, the Department of the Environment and Welsh Office (1989) contend that it should provide a useful framework within which to assess the environmental effects of a project, identify any weaknesses and, perhaps, lead to the adoption of better alternatives which will, eventually, lead to a smoother passage through the planning system.

The content of an environmental statement is laid out in Schedule 3 of SI 1988/1199. In Schedule 3 the information required is divided into two categories:

1) specified information, which is mandatory; and
2) explanation and amplification of the specified information, which is not mandatory.

The specified information is divided into five groups, covering description, environmental data, significant effects, mitigating measures and a non-technical summary:

a) **Description**. This category must include a description of the proposed development, the intended site and the neighbourhood within which the site is situated.
b) **Environmental data**. This must include all raw data collected, including surveys and counts, as well as the analysed data and conclusions. This is a particularly important category since it should include objective evidence of the likely impact of the proposed development.
c) **Significant effects**. Having reached agreement between the developer and planning authority as to which effects or impacts (from **b** above) are significant, the environmental statement must then consider how they will impinge

upon the following: human beings; flora; fauna; soil; water; air; climate; landscape; material assets; and heritage. In addition, the statement must also consider the interaction between any or all of these factors.

d) **Mitigating measures**. Having determined the impact of the significant effects in **c**, the developer must then go on to outline what measures will be taken to mitigate them. There are four principal categories of mitigating measure:
   i)   control – such as the filtering of emissions;
   ii) scaling down – such as screening with trees;
   iii) restoration – such as the replacement of lost woodland or open space; and
   iv) compensation – usually through planning gain, such as the construction of new recreation and community facilities.

e) **Non-technical summary**. As the name suggests, this would comprise a broad summary of all the information contained in **a** to **d** above, highlighting the principal effects of the proposed development and the ways in which these effects can be overcome.

The explanation and amplification of the specified information should be contained in a seven-part classification, covering: physical characteristics; materials and processes; residues and emissions; alternative development; the use of natural resources; the methodology of assessment; and any difficulties encountered in preparing the environmental statement:

a) **Physical characteristics**. This is concerned with both the project itself, the land taken for the development, the land temporarily used during construction and any land taken permanently for mitigating measures such as screening, open space or new rights of way.

b) **Materials and processes**. This category is most appropriate to refining and manufacturing projects, where the nature of the raw materials being delivered to the site and the process of conversion occurring on the site are significant. There is no need, however, to assess the likely hazards or methods of dealing with accidents.

c) **Residues and omissions**. This category should include all wastes associated with the project, regardless of whether they are pollutants or not. Included in this category would be heat, light, vibration, smoke and any spoil or other waste.

d) **Alternative development**. This should include details of other sites, designs or processes that were considered at the feasibility stage but rejected by the developer. Reasons for rejection should be given.

e) **Use of natural resources**. This should be included in all cases, but especially where they are in short supply.

f) **Methodology of assessment**. A general category covering all the techniques used in arriving at the assessment of effects.

g) **Difficulties**. This category should note any weaknesses in the environmental statement caused through difficulties in obtaining or analysing information. This is not to excuse a poor statement, but to alert those reading or evaluating it to the insurmountable problems associated with dubious or unobtainable information.

The environmental statement should then be concluded with a non-technical summary bringing all the findings together in a form accessible to all those wishing to read it or likely to be affected by the development of the proposed project.

Whilst it is clear from this section that the principal developments for which environmental assessment was intended are the large extractive and manufacturing industries, some recreation developments are expressly included, whilst others will have significant environmental effects, if only for the fragile locations in which many of them are set. It should further be recognised that recreation is also an important consideration in the measures taken to mitigate the effects of some developments, suggesting that recreation development can have both positive and negative environmental effects, depending on the nature of the project, its size and location. Since the positive externalities associated with recreation development have been discussed earlier in the book, the remainder of this chapter will deal exclusively with the negative environmental impacts of recreation development.

8.3 The environmental impact of recreation development

It is abundantly clear from the literature that whilst, there is a wide range of environmental impacts associated with sport, recreation and leisure pursuits, little is actually known of how wide-ranging or severe these impacts are. In his work on sport, Bale points out that:

> We know relatively little about, and lack a conceptual framework for looking into, the impact of sport on the natural environment. We ... may note that ecological damage, while not especially significant by the standards of many modern-day activities, can nevertheless be brought about by sports. (1989, 137).

Work on the environmental impact of tourist development is a little more advanced, although the complexity of the issues involved has limited the degree to which it can be accounted for in policy making (Pearce 1985). In one of the early attempts to provide a systematic overview of the environmental impacts associated with tourist development, Cohen (1978) suggests that the principal determinants are: the intensity of site use; the resilience of the ecosystem; the time perspective of the developers; and the transformational character of the development. In a later study, Stroud (1983) found that any latent environmental impacts would be intensified by the scale and quality of a development, whilst some developments were simply located in inappropriate areas, such as wetlands and deserts.

In their work, Pearce (1985) and Cohen (1978) recognise that one of the principal hindrances to their efforts was the piecemeal and isolated nature of much of the research on environmental impacts. Edwards (1989), in concurring with this view, notes that most of the work on the environmental impact of recreation on historic sites has similarly been partial, with little attempt to examine the full range of such impacts, nor to evaluate their cause and effect. This has inevitably meant that it has not been possible to develop any theories or models of the environmental impact of recreation on historic sites. The effect of this is that each operator is faced with determining the likely environmental impacts of different management policies at individual sites, rather than being able to start with a conceptual model of what is likely to occur. This is particularly so for organisations such as the National

Trust, where success in attracting visitors and displaying sites inevitably increases the problems associated with the preservation of those sites.

Indeed, the dilemma faced by organisations such as the National Trust represents one of the central and, to some extent, most insoluable problems associated with recreation and tourism. This is because the basis of much tourism and a significant amount of recreation and sport is associated with environmental factors, whether in the form of historic sites, landscapes or simply open country. As Pearce (1989) indicates, tourists in particular are actually attracted to the more complex and fragile environments, such as coastal zones, alpine areas and sites of historic and cultural interest. In addition, these environmental factors are also important to poor regions or countries in attempting to gain new sources of income:

> ... tourism can sometimes provide a conundrum. A poor country has a natural or cultural attraction and wishes to exploit to the maximum the economic benefits which can flow from it in the short term; foreigners may be concerned about its preservation – but those foreigners are not on the poverty line. But after a few years of successful tourism, the attraction can suffer permanent damage, so depriving future generations not only of the attraction but the possibility it offers to provide tourism flows and economic benefits. (Bodlender 1990, 249).

This problem is characterised by Cohen (1978) as the dilemma between the need to protect the environment for tourism and the need to protect it from tourism. Whilst the former might be a goal for all areas or countries, the latter is highly dependent upon the nature of the tourist destination and the economic and political strength of individual nations. A current example of this is the extent to which local communities can exploit the tourism potential of some of the lesser known islands of the West Indies without allowing it to dominate their culture, as has become the case elsewhere.

In considering the environmental impacts themselves, it is necessary to distinguish between those relating to individual activities and those arising as a result of physical developments associated with recreation and tourism. Whilst there is considerable evidence of both in the literature, the work conducted on individual activities is, understandably, more frequent and comprehensive. It also tends to be more directly related to the physical environment than to people or society. In what is, arguably, the most comprehensive study, Edington and Edington (1986) suggest that whilst it is easy to imagine that activities such as trail biking, skiing and snowmobiling have an effect on the environment, it should also be recognised that some apparently harmless, sensitive activities, such as birdwatching, can be equally damaging if the participants are not aware of the effect that they can have on the environment. In the case of birdwatching this could involve the disruption of the behavioural and ecological relationships of the birds that are being watched.

There are many examples of the types of environmental impact of recreation, sport and tourist activities contained in the literature. These range from the general overviews of the ecological impact of recreation and tourism contained in Edington and Edington (1986) and Pigram (1983), to the more specific relationship between sport, recreation and nature conservation to be found in Sidaway (1988), the environmental impact of sports activities (Bale 1989) and the impact of visitors on historic sites (Edwards 1989). Examples of these various impacts, drawn from the authors cited, are contained in Table 8.1.

Table 8.1 Examples of the environmental impacts of sport, recreation and tourism

Activity	Environmental effect
Climbing	Disturbance of cliff-nesting birds; damage to mountain vegetation.
Water recreation	Disturbance of water birds; damage to banks by water.
Horseriding	Damage to vegetation.
Recreational vehicles	Damage to desert ecosystems; Damage to coastal sand dunes.
Observing wildlife	Disturbing natural relationships, especially by artificial feeding.
Hunting and fishing	Disruption of natural communities by the introduction of game animals and fish; intensification of hunting caused by the tourist souvenir trade.
Sight-seeing	Trampling damage to vegetation and soils; natural vegetation replaced by more recreation-tolerant species; environmental changes at historical and archaeological sites.
Caving	Disturbance to hibernating bats.
Walking	Disturbance to ground-nesting birds.
Orienteering	Disturbance to birds during nesting season.
Canal restoration and use	Disturbance of a rich variety of aquatic life.
Motor sports (land)	Chemical pollution; noise.
Motor sports (water)	Oil and petrol spillage; noise.
Alpine skiing	Damage to woodland, grass and soil.
Cross-country skiing	Damage to undergrowth.
Recreation-related travel	Air pollution; noise.

(Source: from Bale 1989, Edington and Edington 1986, Pigram 1983, Vasallo and Delalande 1980 and Sidaway 1988).

It is clear from Table 8.1 that, whilst most recreation activities impose some form of impact on the environment, the nature, intensity and gravity of these impacts vary considerably. Equally, as our knowledge of environmental impacts improves, so concern has shifted from a concentration on trampling and other vegetative impacts to concern over the disturbance of fauna. Added to this is the increasing recognition that the level of recreational use can be as important as the type of use in determining the environmental impact of a given activity. Thus, with respect to the interaction of water recreation and over-wintering waterfowl, problems have only recently begun to emerge as water recreation has ceased to be solely a summer activity, thereby encroaching on the winter use of water by birds (Sidaway 1988). Because of this, a former harmony between the participants and the birds has been replaced by conflict, particularly on the crowded waters of the South coast of England, leading to opposition to any measures designed to increase the number of boats or reduce boating congestion.

Equally, the traditional pursuits of hunting and fishing have probably come under scrutiny as much as a result of their increasing popularity as any recognition of the ideological implications of supporting, or allowing, them to continue. Thus, the recent majority vote by the members of the National Trust to ban stag hunting on the Trust's land certainly came about as a result of increased knowledge of such activities. Furthermore, the fact that the ban was not implemented has merely added extra weight to the belief of many that recreational hunting is increasingly indefensible:

> If the same criteria are applied to recreational hunting as to other recreational pursuits, namely that they should be practised with minimal environmental damage and interference with other interests, then the record for hunting is generally a poor one. The continued failure to take a responsible attitude to introductions of game species and to unjustified 'control' of predators, are major grounds for criticism, as also is the tardiness with which the lead-contamination problem has been tackled. (Edington and Edington 1986, 75).

The damage caused by people to both formal and informal recreation sites is well known, particularly in terms of pollution in areas such as the Solent on the South coast of England, the erosion of vegetation and soils in coastal areas of Lincolnshire and Norfolk and the degradation of buildings and their furnishings and fittings, as has been the case in cities such as Venice. It is also in this context that the relationship between the numbers of visitors or users and the resulting impacts is most clearly demonstrated. Indeed, in the most extreme examples this has led to the enforced closure of the site or attraction, as was the case with the Lascaux Caves in the Dordogne, France, following irreversible damage to the cave paintings. A similar fate could still befall Stonehenge in England unless a new means of presentation and control can soon be found, whilst the Major Oak in Sherwood Forest is now so badly damaged that its future is most doubtful. Faced with problems of this magnitude, Edwards (1989) suggests that one of the most pressing needs is for studies of visitor management in an attempt to reconcile the conflict between the need to attract visitors in order to generate income for preservation works, and the environmental impact of those visitors.

The activity of canal restoration and use included in Table 8.1 has highlighted some of the most unlikely debates over recreation and the environment. As canals became disused and dilapidated they contributed very little to the environment, in either positive or negative terms. The determination of enthusiasts to restore them, primarily for recreational uses, tended therefore to generate a positive response from all concerned. There was even a tacit recognition that opening up such waterways could enhance the wildlife value of such areas, in addition to their amenity and recreational values. However, the result of some such restoration projects, such as the reopening of the Basingstoke Canal, have generated heated debates between conservationists and the restorers on the grounds that the rich diversity of aquatic life revealed by the restoration could be destroyed by the intensity of recreational use envisaged by the restorers. The notification of the Basingstoke Canal as a Site of Special Scientific Interest by the Nature Conservancy Council (NCC) has brought this whole issue to a head:

> The NCC argues for quiet use, including canoes and rowing boats, and seeks to limit the number of motorised craft at about 1,000 boat-movements per year. This proposal has antagonised the SHCS (*Surrey and Hampshire Canal Society*), in particular, as use by motorised boats has always been a prime purpose of restoration. (Sidaway 1988, iii).

In terms of the impact of recreational activities on the environment, therefore, it has been shown that, whilst not especially significant when compared to factors such as industrial pollution, there is a range of impacts that cannot be ignored, since the gravity of those impacts is strongly related to levels of use; and recreation and tourism are amongst the fastest growing industries in the world (Bale 1989).

In contrast to the ubiquitous nature of recreation participation, actual developments associated with recreation and tourism are relatively limited and infrequent. Their impact can, however, be immense, particularly in terms of the visual intrusion of new development into unspoilt landscape (Fortlage 1990). Added to this is the undoubted impact of travel, with its associated infrastructure of roads (particularly new and widened motorways), railways and air and sea ports, all of which provide a constant reminder of the ways in which the environment can be modified to suit the needs and wants of human beings.

In his work on tourist development, Pearce (1989) delineates three major sources of environmental stress related to the development of tourist facilities: the permanent restructuring of the environment by major developments; the generation of new or increased waste residuals; and the effect of tourist developments on population dynamics. Although specifically applied to tourist developments, these sources of stress are equally applicable to any form of sport or recreation development. The greatest single ecological effect associated with such development is the replacement of the natural ecosystem with an artificial one (Bale 1989). Examples of this can be found everywhere from a local sports centre or stadium to the coastal ribbon development of southern France, northern Belgium and parts of Spain.

Apart from the loss of the ecosystem, the principal impacts of urban sports and recreation facilities concern increased congestion, particularly at the end of events when all the spectators are attempting to leave at once, increased noise, increased pollution from transport associated with those spectators and, when involving soccer in the UK, increased levels of violence and vandalism. In terms of property itself, some of the environmental impacts of sports stadia are passed on to the community in the form of lower property prices in areas adjacent to a stadium (Bale 1989). This is in contrast to higher property prices for areas adjacent to golf courses, parks and other areas perceived to be of high amenity value.

The principal environmental impacts of coastal development include land and sea pollution, loss of habitat, changes in shoreline configuration, loss of sediment in some places accompanied by siltation in others, increased coastal protection needs and, potentially, access problems. The perceived degradation of coastal areas by tourism and recreation development has become so great in some areas of the UK that the National Trust has initiated the Enterprise Neptune scheme with the aim of acquiring and then protecting particularly fragile areas of coast from such development. Similarly, the Countryside Commission's Heritage Coast initiative has much the same objectives, but without the wish to actually acquire the land.

Recreation development near inland water is another source of potential environmental stress. Apart from habitat loss, pollution from people and motorised craft and the visual intrusion of many such developments, there is growing evidence of inland, as well as coastal, waters such as the Norfolk Broads beginning to suffer from eutrophication, with the consequent loss of aquatic life. Indeed, coral is being killed by algae in the tropics, whilst the whole nature of aquatic life is being altered in more temperate climes.

There is also growing evidence that the commercial potential of some forms of recreation and tourist activity is beginning to influence the actions of organisations not primarily involved in recreation provision. This is particularly so of the recently privatised water companies in England and Wales which, as public monopolies, did so little for recreation provision. Now charged with commercial as well as water supply objectives, these companies are seeking to exploit the potential attraction of water as the basis for recreation development, even where the water supply arguments are not strong. In one particular case reported recently (Dunn 1991), South West Water plc have been accused of selecting a particular valley for a new reservoir on the basis of its recreational, rather than water supply, potential. Whilst the objectors are questioning both the actual need for the reservoir, as well as its siting, the company asserts that a new reservoir is required and that, in addition, its duties under the Water Act 1989 mean that it must take recreational potential into account in its location decision, but that any development will be environmentally sensitive. The question remains, however, of how the loss of a 55 acre valley equates to the commercial gain of a new recreation development, even if that development is sympathetic to the local environment. Another, less contentious example, is the construction of the Queen's Valley reservoir in Jersey. This was formed by flooding a valley to augment Jersey's high holiday season water demand whilst also retaining the amenity and enhancing the recreation potential of the valley (Seeley 1992).

The second area of stress discussed by Pearce (1989) is the generation of waste residuals. The single most important problem within this category is the pollution of water by the discharge of inadequately treated effluent. Since the same area of water that forms the focus of a recreation or tourist destination is convenient for effluent disposal, few operators have resisted the temptation to discharge untreated or partially treated effluents into the water. Thus, whilst McMillan Scott (1990) and others protest that the tourist industry, in particular, is reliant upon looking after its own environment since it is dependent upon it, the vast majority of beaches and many inland water courses are becoming dangerously polluted partly through the action, or inaction, of the tourist industry. However, the tourist industry itself argues that much of the problem is caused by inadequate investment in infrastructure by government, particularly in areas where the development of tourist accommodation and associated facilities is expanding rapidly (Pearce 1989). Also associated with major tourist developments can be problems of hypersalination of water where desalination plants have to be employed to obtain drinking water, the lowering of water tables where bore holes are sunk for the same purpose and soil erosion where trees are cut to facilitate development and provide fuel for the new tourists (Edington and Edington 1986). Finally, there is a growing body of evidence that air pollution in tourist and recreation areas, caused by the proliferation of vehicles bringing visitors, is actually reducing the quality of air to less than that found in many cities (Pearce 1989).

The third major aspect of environmental stress caused by recreation and tourist development is the effect on population dynamics. The congestion associated with sporting and entertainment venues has already been discussed above. However, the whole culture of tourist areas can be threatened by the annual influx of tourists, regardless of whether it is a resort on the Greek Islands or the coast of Britain. Such marked seasonal variations in population numbers as is experienced at many coastal resorts also puts an unnatural demand on natural resources, such as water and energy, meaning that many areas are faced with planning for peak times, but asking permanent residents to shoulder a disproportionate share of the cost.

Whilst it has certainly been the case that much of the recreation and tourism industry has turned a blind eye to the environmental impacts of its activities, especially in some of the coastal resorts of the Mediterranean and other tourism honeypots, some ameliorating steps have been taken. Indeed, one of the most important concepts in understanding and limiting the development of any given site, namely carrying capacity, was first discussed nearly 30 years ago (Wagar 1963). The idea of carrying capacity is that every site has a maximum capacity that cannot be exceeded without damage to, or modification of, its natural attributes. Although developed specifically for use in North American parks, the concepts are equally applicable to any recreation or tourism development.

Patmore (1983) suggests that there are four principal types of carrying capacity:

a) **Physical capacity**: the maximum number of visitors a site can accommodate without physical deterioration;
b) **Ecological capacity**: the level of use a site can endure without experiencing irreversible ecological damage;
c) **Perceptual capacity**: the maximum level of use that can be permitted without unduly impairing people's recreation experience; and
d) **Economic capacity**: the minimum level of use necessary to generate an acceptable financial return.

In order to be fully sustainable in the long term, no recreation or tourism development should exceed any of the first three capacity levels, whilst endeavouring to attain the minimum level of economic capacity.

Whilst the concept of capacity is relatively straightforward, the determination of appropriate levels at any given site is quite the opposite. As Edwards (1989) comments, few studies have been undertaken into the relationship between visitor numbers, visitor types and the impact that they have on particular sites. This inevitably means that most operators are faced with setting capacity levels on the wholly inadequate basis of intuition and then adjusting them in the light of experience. The need for comprehensive studies of capacity selection cannot, therefore, be over-emphasised.

The concept of carrying capacity is further limited by its concentration on individual sites. Firstly, this means that it is not appropriate for many open or informal areas of access, whilst it also fails to take into account any environmental impacts that extend beyond the boundaries of the enterprise or development. In response to these shortcomings, a wider approach has begun to find favour with policy makers, particularly in the field of tourism, in the form of sustainable tourism development:

> Tourism's environmental track record is not impressive. But community-led sustainable tourism development (STD) may change this. The concept of STD goes further than the fashionable green approach, dealing with the management of all resources. It aims to create a system where economic, social and aesthetic needs can be fulfilled, while maintaining cultural integrity, ecological balance, and life support systems. STD embraces heritage and culture and is entirely based on long-term objectives. The vital ingredient in the equation is the involvement of local people in planning, developing and managing tourism and leisure activities. (Stevens 1990, 64).

The interest in environmentally-aware tourism and recreation development has grown fast, with many of the leading private and public sector agencies now actively involved in improving their environmental performance (Gordon 1991). At present, it is probably the case that most initiatives have been centred on community participation in planning, development and management, rather than ecological sustainability. However, there is a strong argument that this alone is enough to promote a much greater degree of environmental awareness by both operators and visitors.

Unfortunately, notable absentees from this shift towards environmental sustainability have been the agencies involved in planning recreation and tourism development in Britain. Whilst Britain's Overseas Development Agency has adopted the philosophy of sustainability in its development work, the predominant philosophy in Britain remains that of the market, with little apparent place for community-led sustainable tourism development (Stevens 1990). Without a longer-term view, however, the prospects for environmentally sound tourism development are not promising:

> Tourism is ... involving more and more people and becoming a virtual mass phenomenon whose uncontrolled expansion can be seriously damaging for the environment. Effective organisation and regulations are thus essential; hence it is vital for governments to know what the future prospects for tourism are so that they can adopt long-term policies in order to intensify, modify or check trends in tourism growth in line with environmental requirements. (Vasallo and Delalande 1980, 41).

8.4 The impact of statutory environmental assessment on recreation development in Britain

Although relatively few forms of recreation development are formally included in SI 1988/1199 and although there has been a short space of time since it came into force, a range of recreation developments has been considered suitable for environmental statements. Of the 250 or so developments requiring environmental statements that were notified to the Department of the Environment up to June 1990, some 33 were wholly related to recreation and tourism, whilst a further 20 had a substantial element of recreation in them. Since a further 75 applications concerned either the extractive or waste disposal industries, recreation facilities comprised about one-third of all those actual developments for which an environmental statement was required. The main categories of recreation development were marinas, holiday villages, hotel complexes, ski lifts and racing circuits. In addition, there was a range of one-off development proposals, comprising theme parks, film studios and tramways. The majority of the 20 proposals with a substantial element of recreation included those related to shopping and leisure, retail and hotels, offices and hotels and a sports complex within a town centre redevelopment programme.

The surprisingly large proportion of proposals concerned with recreation was partly as a result of the buoyant market conditions at the time, with many developers feeling that the age of leisure was at last arriving, and partly as a result of the sensitive location of many of the proposed developments (Gosling 1990). Indeed, it would appear that most of the developers involved fully accepted the need for the environmental impact of their proposals to be assessed, since none have so far challenged the local planning authority's request for an environmental statement.

One of the principal problems facing many recreation developers has been whether their proposals will actually require environmental statements, given that there are no thresholds or guidelines in the legislation (Day and Davis 1990). This is exacerbated by the relatively small size of many of the development companies involved. Since they are unlikely to have the staff to produce a comprehensive and technically valid statement, and since many of their applications will be in outline only, the process of preparing a statement can be both time-consuming and costly. This has led, in some cases, to inadequate or inappropriate statements being produced, often with an over-emphasis on the qualitative aspects of the development, at the expense of the quantitative ones (Gosling 1990). Alternatively, it has led to the production of highly elaborate comprehensive statements that have given opposition groups plenty of ammunition with which to oppose developments that they see as undesirable.

Other problems that have been highlighted, although not necessarily related to recreation development, include the inevitable delay in the development process caused by the need to prepare an environmental statement and allow time for it to be assessed by the planning authority and other interested parties, the clarification of the acceptable contents of an environmental statement and the problem of recognising the full impact of a new development in a greenfield location, where the visual intrusion may be difficult to judge at such an early stage of the development.

However, the widespread use of environmental statements in Britain is at such an early stage that it is, as yet, difficult to know what the longer-term will hold. In the USA, where they have been in use for longer, the majority are passed without comment. This is partly due to the sheer weight of numbers involved and partly to the agencies concerned becoming practised at preparing the statements:

> They have learned to avoid, as far as possible, contentious issues and areas, have been careful to conform to the letter if not the spirit of the legislation and have employed the assessment process to defuse potential conflicts. (Rees 1990, 351).

8.5 Conclusions

It is certain that there has been an important reappraisal of the role of development and its relationship to the environment over the past few years. As the case studies in Chapter 9 will illustrate, the role of environmental assessment has now moved from the periphery to the centre of the development process, both as a means of identifying the likely impact of the development and of providing a more formal channel for public involvement in statutory planning. However, it is important not to over-estimate the significance of this shift for, as Rees points out with respect to the USA, environmental assessment 'merely added a new procedure which had to be undertaken before formal decisions were made, it did not significantly alter the decision system *per se*' (1990, 349). This is not, however, to undermine the benefits that can be reaped from such a system, both in terms of attempting to quantify the impacts and in beginning to create or foster a culture of valuing the environment.

Nevertheless, there is still much to be done. Quite apart from the operational problems associated with environmental statements in the British planning legislation, there is still a feeling that the biggest benefit to be gained by developers in agreeing to prepare environmental statements is an improvement in their public image (Day and Davis 1990). Equally, the presumptions within the legislation in favour of development, and of associating environmental damage with new development, remain to be challenged.

Before this can be achieved, however, there is much ground to be covered in co-ordinating a cohesive response to the environmental impact of much development:

> Some of the fiercest debates in the environmental literature are between preserva-
> tionists and conservationists. ... If, for example, there is a belief in the rights of
> wildlife or the rights of living things in general, it is difficult to square these,
> superficially anyway, with the idea of managing a natural resource for human
> benefit. Similarly, many preservationists think that conservation as a compromise
> between development and preservation gives too much ground – it is the 'thin end
> of the wedge' in the fight to protect natural environments. In many cases, the
> conservation option does not really arise: either a given habitat is preserved
> because it is the minimum critical area needed for species survival, or it is
> destroyed for development. (Pearce and Turner 1990, 311–2).

Quite apart from the philosophical gulf that can separate different parties in the development process, Sidaway (1988) found that all too often there was a basic lack of understanding between them. This was particularly so in the failure to understand the relationships between different species of fauna and flora on the part of developers; and the unwillingness to respond to dynamic situations, such as the colonisation by, or decline of, a char-acteristic species on the part of conservationists. These same findings, although related to historic sites rather than nature conservation, led Edwards (1989) to the conclusion that much more work is necessary on the management of sites, in stark contrast to the emphasis put on development by the legislative imposition of environmental assessment.

However, in overall terms it has to be concluded that the harm caused by recreation and tourism is, for the most part, relatively minor when compared to the major sources of pollution and environmental degradation. This does not mean that the interest shown in the assessment of the environ-mental impact of recreation and tourism development is misplaced, or that the local impact of such development cannot, on occasions, be severe. It is, however, to point out that such development can bring many benefits to host communities, not to mention tourists and recreation participants, but that for those benefits to outweigh the environmental costs there must be a formal recognition of the need to achieve a lasting balance:

> Planning at all levels can help to increase the economic, social and environmental benefits which tourism may bring and at the same time reduce the associated costs. This can more readily be achieved where a broader approach is adopted, where goals and objectives are clearly defined and related to local, regional and national needs, where sound resource evaluation is complemented by an analysis of tourist demand and marketing and where a legal and administrative framework allows plans to be not only devised but also implemented. (Pearce 1989, 278).

9 The Interaction Between Recreation Development Proposals and the Planning System

9.1 Introduction

The intention throughout this book has been to produce a text that deals with the practicalities of the leisure development process, albeit in a perceptual rather than sequential approach. Reference has therefore been made to many actual examples, both in the United Kingdom and elsewhere. However, in only referring to illustrative examples where appropriate, it has not been possible to build up a comprehensive picture of how the leisure development process moves, or fails to move, from inception through to operation, as outlined in the book. This chapter will attempt to fill that gap by describing four case studies from southern England which illustrate how the process has developed over the last twenty years.

Whilst there is a well established land market for many uses, whether agricultural, residential or commercial, no such market can be clearly delineated for leisure development. This inevitably means that for the most part, leisure developments are confined to play a residual role in changing land use patterns. In urban areas, therefore, land for leisure developments is often released as a result of other, non-leisure development, on land already owned by a local authority, or in an attempt to conserve an historic aspect of a town or city.

In rural areas where there is rather less development activity, there are less obvious opportunities for any type of leisure development. Conversion of land to leisure usage is, therefore, often contentious since it is likely to have displaced a more accepted or expected use of that land. One such opportunity to have emerged in many rural areas over the last two or three decades is the use of worked-out gravel pits for a variety of leisure activities. Although most planning permissions for mineral extraction have required the land to be returned to its former use, usually agriculture; many wet pits (where the water table is above the level of the pit floor, causing natural flooding of the pit) have been used for fishing, sailing and canoeing, even before extraction is complete. Whilst these types of uses have rarely afforded the pit owner a commercially viable enterprise in the long-term, the growing demand for leisure facilities, particularly in the south of England, has provided the potential for many commercial leisure development proposals. Two of these latter types of proposal, Thorpe Park, west of London and Somerford Keynes holiday village, near Cirencester, Gloucestershire, form part of this chapter (see Figure 9.1).

At the same time that the leisure use of worked-out gravel pits has become more common, all mineral workings have come under increasing legislative control. In particular, there is a growing sensitivity to both the extraction and the restoration of the pits, with aggregate producers' restoration records increasingly being considered in any application for permission to extract (RMC Group plc 1987). In particular, the Town and Country Planning (Minerals) Act 1981 imposes aftercare conditions for five years from the final reinstatement of the site, so providing an incentive to both restore the site properly and ensure its future management.

Faced with these conditions, many aggregate producers have begun to consider the scope for leisure development, both as a means of ensuring adequate restoration and as a means of providing a commercial return to cover the inevitable cost of aftercare. It is, furthermore, no surprise that aggregate producers see leisure development as a means of demonstrating their public spirit, particularly where they include provision for wildlife habitat, but also in meeting the public's growing demand for leisure facilities (Oliver 1984). The result of this is that there has been a significant increase in proposals for developing worked-out pits, whether for leisure, as in the case of Thorpe Park and Somerford Keynes, or sport, as in the case of the National Water Sports Centre at Holme Pierrepont, Nottingham, or for other types of development, often residential in nature. Many of these proposals have come from the aggregate producers themselves, using subsidiary companies such as Leisure Sport Ltd (owned by RMC Group plc) or ARC Properties Ltd (owned by Amey Roadstone Company), or from outside operators, such as Park Hall Leisure plc, a subsidiary of Granada Group plc and, latterly, Lakewoods Ltd, a joint venture between Granada and developers John Laing, which acquire land from the aggregate producers and take on their responsibilities for restoration and aftercare.

The two development proposals to be considered in this part of the chapter are separated in time by some twenty years. When first proposed at the end of the 1960s, Thorpe Park was hailed as Britain's first theme park (Anon 1980). Although not opened until the end of the 1970s, it has since built a reputation as one of Britain's leading leisure attractions, regularly attracting over one million visitors per annum with a wide range of exhibits and rides for the whole family. Still very much at the planning stage, Somerford Keynes holiday village represents an equally unique form of attraction, comprising an activity-based, self-contained village for both short stay and full holidays. Whilst being very different in concept, the uniqueness of both Thorpe Park and Somerford Keynes has tested the ability of the planning system to react to innovation and change. Their stories indicate that whilst much has changed, in both the planning system and the methods of application, the essential elements of the innovative role of the developer and the regulatory role of the planning system have remained unaltered, with neither side of the equation seemingly any closer to understanding or anticipating the demands of the other.

In contrast, the development of leisure facilities in urban areas has been marked not by planning issues as much as by the high cost of building land and the sometimes uneasy relationship between public and private providers. Two areas in which much leisure development has occurred have been public sector sports and leisure centres, and the conversion and new leisure use of otherwise redundant historic buildings. In both these cases, such developments or alterations are generally welcomed; often a very different scenario from proposed development in the countryside.

This is not to suggest, however, that proposed leisure developments in urban areas are without problems. Whilst there is rarely too much opposition to the concept of public sector provision, most tends to occur on land already owned by the local authority, meaning that the location might not be ideal for the proposed facility, whilst any current use of the site, often involving informal access and recreation, will be forfeited. This was certainly the case with the first of the urban case studies, Guildford Borough Council's new leisure centre built on part of Stoke Park, a well used public park near the town centre (see Figure 9.1).

Equally contentious as the development of public open space is the re-use of historic buildings. Whilst a wish to control such alterations in urban areas might seem somewhat hypocritical given the legacy of refurbished historic

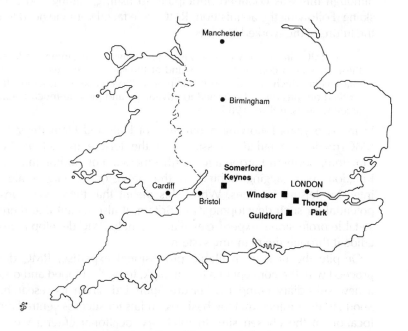

Figure 9.1 The location of the case study sites

buildings used for retail purposes in most town and city centres, it is certainly apparent that local authorities do wish to conserve any remaining buildings. The major problem that they face in central areas is, however, attempting to find a use that will not irretrievably alter the facade and character of the building. This almost certainly means a presumption against retail, residential and, often, office uses, whilst very few industrial or warehouse uses would be acceptable in a town centre, even if occupiers prepared to pay town centre prices could be found. Leisure therefore becomes a most attractive proposition for the local authority, although it may be less so for the landowner due to the lower rent or purchase price payable by a leisure enterprise, when compared to retail, residential and office developments. In some instances, such as the subject of the fourth case study, Royalty and Empire at Windsor (see Figure 9.1), these factors can converge so that leisure is seen to be the most appropriate option by the landowner and the local authority, even in an already overcrowded tourist centre such as Windsor.

9.2 Thorpe Park

Thorpe Park's history as a commercial recreation site began in 1968 when the Ready Mixed Concrete Company (RMC) acquired Hall and Ham River Company, together with lots of its worked-out gravel pits in the Thames Valley. There was already some recreational use of many of the pits, although this was confined principally to fishing, sailing and some water skiing. Following the acquisition, RMC were faced with the need to consider the future of the worked-out pits:

> ... the RMC Group plc was anxious to find ways of developing at least some of these sites on a commercial basis and at the same time create areas for public enjoyment which would also be helpful to the company in demonstrating that worked-out pits could be turned to advantage and provide much needed public amenity areas. (Oliver 1985, 171).

In the two years following acquisition of Hall and Ham River Company, RMC made a detailed assessment of the leisure market in the United Kingdom, northern Europe and North America, concluding that the United Kingdom was lagging behind in the provision of large-scale, capital-intensive leisure facilities. Whilst this meant that there was considerable potential for such developments in Britain, it also meant that there was no suitable professional expertise available to help RMC develop a proposal or guide it through the planning system.

Despite the lack of available professional expertise, RMC decided to proceed with the concept of a theme park, to be developed and operated by a new subsidiary company, Leisure Sport Ltd. From its research, Leisure Sport Ltd determined that the basic essentials for success centred around the location of the chosen site. In total, six locational criteria were used to determine the most appropriate site to develop:

1) the site should be near major centres of population;
2) it should be close to motorways or major trunk routes;
3) there should be a good network of local roads leading to the site;
4) the surroundings of the site should enhance the development itself;
5) if possible the site should be in an established tourist area where the planning authority and local inhabitants would recognise the benefits of the development;

and

6) public transport to the site would be a considerable bonus.
(see Oliver 1985).

After considering all of the available pits, those around the village of Thorpe, some 20 miles (32 km) west of central London, were felt to be the most suitable, although being within the London Green Belt. In terms of the six criteria, the Thorpe pits matched three of the first four, being close to London and the Surrey and Berkshire commuter belt, being next to the M3 motorway and close to the proposed route of the M25 motorway (see Figure 9.2) and being surrounded by a wooded, semi-rural environment (Catliff 1983). Whilst the network of local roads was not ideal, Leisure Sport Ltd were able to satisfy the planning authority that the network could be improved sufficiently by the development of a new roundabout, at the company's expense, diverting traffic away from Thorpe village. Also, being close to London, the possibility of arranging public transport to the site was not considered a problem. This meant that of the six initial criteria, five could be met by the Thorpe site, whilst proximity to London, itself a tourist centre,

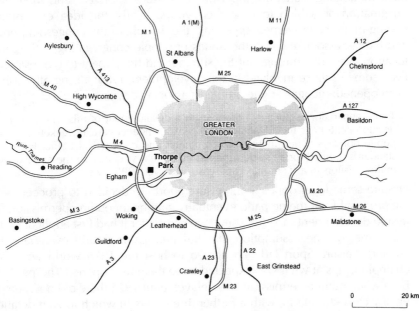

Figure 9.2   The location of Thorpe Park

meant that Leisure Sport Ltd felt entitled to believe that the local authority would appreciate the potential benefits to the locality of a new recreation development.

Having decided, by early 1970, that the Thorpe site was suitable for a leisure development, Leisure Sport Ltd approached the planning authority to seek its view on a change of the reinstatement provisions from infilling, which had already started on 60 acres (24 ha) of the site, to leisure. The county planning officers, as representatives of the minerals planning authority, were generally in favour of the proposal, although stating that they would only be prepared to grant a change in the permission if there was no local opposition. Armed with this encouragement Leisure Sport Ltd set up a series of meetings with local residents to explain the concepts and benefits of the proposal. Quite apart from the increased prosperity and job opportunities that the new park would generate for the area, the company were keen to impress upon the residents that if the full infilling requirements of the outstanding planning permission were complied with, residents would have to endure twenty to twenty five years of tipping, probably followed by severe pressure for residential development. By agreeing to the leisure proposal, complete with new road scheme, the residents would not only avoid the tipping, but would also gain a carefully landscaped and protected environment. Leisure Sport Ltd further undertook to ensure that only low-grade activities would be sited close to the existing residential areas (Catliff 1983 and Oliver 1985).

Quite predictably there was some concern and opposition to the proposal. There were no similar sites already operating in the country, meaning that local inhabitants had only the word of Leisure Sport Ltd and their own imagination on which to base their views. Whilst the idea of avoiding tipping might have seemed appealing, the thought of the congestion, noise and danger associated with thousands of people congregating at the same location every day throughout the summer did not appeal to the residents. Even some of those involved in the development, rationalising after it had been opened, recognised the problems associated with traffic:

> Traffic generation is always a problem in a development of this type ... Leisure developments have the advantages that peak attendances are at weekends and on public holidays, when there is no commuter and commercial traffic. However, standards applied today seldom permit additional volumes on the scale of one million visitors in a six month Park season. (Oliver 1985, 173).

Despite some local opposition, Leisure Sport Ltd decided to proceed with the proposal for a theme park, eventually gaining an outline permission for leisure development in 1972, some two years after it had first applied. In the meantime, on the assumption that full permission would eventually be granted, Leisure Sport Ltd had agreed to host the 1975 world water ski championships at what was expected to be the newly opened Thorpe Park. Even when outline permission was delayed until 1972 there did not seem to be any cause for alarm, with a further three years in which to gain detailed permissions and develop the facilities.

Carrying on with their preferred policy of permission through negotiation, Leisure Sport Ltd began to apply for the detailed permissions for individual parts of the site. Whilst some were successful, a number of important elements of the original proposals had to be abandoned as their permissions were rejected. These included an ice rink, amphitheatre, water flumes and a number of childrens' rides. However, progress was still being made, if somewhat slowly, when the reorganisation of local government occurred in 1974. Following the reorganisation, with planning powers moved from the county to the district authority, the local opposition to the proposed development gained momentum. This ensured that by 1975, and the arrival of the contestants for the world water ski championships, there was no park and few facilities.

The championships went ahead, nevertheless, attracting much favourable comment about the concept, if not the reality of the park. One of the most vociferous supporters of the park was the then Minister of Sport, Dennis Howell, MP, who declared that the park represented a major fulfilment of government policy (Oliver 1984). Upon hearing of the local planning difficulties, Mr Howell arranged a meeting with the local planning commit-tee to try and resolve the difficulties that had developed between Leisure Sport Ltd and the local residents. The outcome of this intervention was mixed, for whilst enough detailed permissions were eventually forthcoming to make development viable, the authority did issue an Article 4 Direction limiting the future use of the site. The Direction remains in place today.

By 1976, some four years after the granting of outline permission and six years since the first application for permission to develop was made, Leisure Sport Ltd was ready to begin development of the site. In order to get to that position there had been nearly 150 planning applications, three appeals and a compulsory purchase order. In addition to the time delay, Leisure Sport Ltd had also spent a great deal of money, prompting one commentator to state that:

> Planning for the growth of leisure activities has official Government approval, but the Thorpe Park experience is perhaps not likely to encourage imitation. It could only have been attempted by an organisation with considerable resources and enterprise and immense patience. (Catliff 1983, 489).

Whilst some construction had been undertaken on the site since 1971, the majority of the work was completed in the period from 1976 to May 1979, when it was officially opened. In its final configuration Thorpe Park consisted of a number of themed areas, including catering and entertain-ment at the Mountbatten Pavilion, a childrens' rides area, a range of educational exhibitions, the water ski lakes and some nature reserve areas, all arranged around a series of interconnecting lakes. A demonstration farm was added in 1982. Attendances in the first year of opening were 300 000, rising to 600 000 by 1982 and over one million by 1984. Since that time management has sought to maintain visitor numbers at between 1.1 and 1.3 million per annum, which is felt to be a comfortable maximum for the site.

Following the eventual opening in 1979, the existing management was replaced with a new team in 1980. Apart from the actual operation of the park, the primary task of the new team was to establish an improved relationship with the planning authority. This was done over a period of time, culminating in an agreed scheme that would limit future development to the central core of the park, well away from any local residents.

However, the Green Belt designation of the land meant that each future planning application would continue to be treated as a change of use, rather than as an extension of the leisure use of the site. In 1985, therefore, a successful application was made to change the designated use from Green Belt to leisure. Following this legitimisation of the leisure usage it was felt that the future development of the park could proceed with rather more confidence than previously.

With so many visitors and being so close to a major population centre it is inevitable that there will be a lot of return visitors. Indeed, the management has sought to encourage this as a means of attempting to guarantee, or at least predict, the annual visitor numbers. However, in encouraging return visits, management is also committed to a constant programme of development and updating the site. However, in so doing Leisure Sport Ltd has been accused of changing the original concept, thereby deceiving people into agreeing to things that they were not aware of. In its defence, Leisure Sport Ltd has argued that a change of image is inevitable in a project that was already ten years old before it was opened, but that since agreeing the development zones and the leisure designation, everyone has at least been aware that Thorpe Park will continue to be a well managed leisure attraction operated by responsible management. However, it was not until as recently as 1989 that permission was granted for the development of some major rides, such as the 47 feet high log flume, that had originally been rejected.

In many ways it is evident that the proposals for Thorpe Park were simply too early in the evolution of the leisure planning process. At a time when traffic congestion, particularly at holiday resorts and tourist destinations, was becoming a serious problem, and when aggregate producers such as RMC did not have a good name in the restoration of mineral workings, it is, perhaps, understandable that the local residents and planning authority were wary of such large scale proposals. It is also questionable just what level of information was available about the possible effects of the development, either to Leisure Sport Ltd or to the local people. In retrospect, therefore, the incremental approach to the planning system adopted by the company may have had quite the opposite effect to that intended, as Oliver postulates:

> There is no doubt that much of the planning delay could have been avoided if the company had submitted a detailed application at the outset, and then deemed a planning refusal and taken the matter to appeal and public inquiry. Local people would have been given the opportunity of having their say and the company would have stood every chance of obtaining a permission for the development in its original form.

The company believed so much in the Park development and in the planning gain to the planning authority of preserving the site from use as a rubbish tip, that it decided to attempt a negotiated permission rather than have recourse to the appeal procedure. (1985, 174).

Quite what the position would have been if the company had adopted a more adversarial stance will never be known; it might have been an outright rejection, ending its ambitions before it had had a chance to demonstrate its faith in the development of a theme park. What is clear is that Leisure Sport Ltd have had to remain attentive to the demands of the planning authority throughout the life of the park. Indeed, the Article 4 Direction has ensured that the planning authority remains aware of, and has to give approval for, any developments intended for the park. However, what is equally clear is that Leisure Sport Ltd and the planning authority have now reached a much more amicable and responsible position, where regular consultation is proving beneficial for all the parties concerned.

9.3 Somerford Keynes Holiday Village

Coming nearly twenty years after the original application to develop Thorpe Park, the proposals for a holiday village at Somerford Keynes are equally as much a perceived response to national government policy and public demand. Unlike Thorpe Park, however, the proposal at Somerford Keynes is for the further development of the Cotswold Water Park, an established recreation and conservation area based on worked-out gravel pits south of Cirencester (see Figure 9.3). Although originally subject to the same type of reinstatement clauses as the pits at Thorpe, it had been recognised in the statutory plans for the area that the pits around Cirencester are well suited to leisure uses, in part to make up for a lack of facilities in the locality. Already

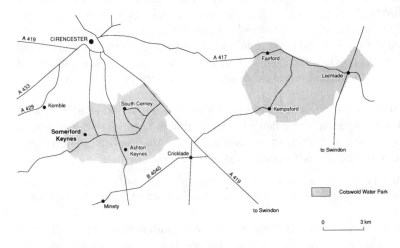

Figure 9.3 The location of the proposed Somerford Keynes holiday village

in the area are a country park and numerous water sport facilities, even though there is still plenty of gravel yet to extract. However, the environment of the area is badly degraded by the mineral extraction operations, leaving a flat, featureless landscape with little more than water, roads and scrub vegetation.

It was in 1987 that the Granada Group plc, through its subsidiary company Park Hall Leisure plc, started looking for a suitable site for a new holiday village concept. Although a similar village had been developed in Sherwood Forest by Center Parcs, the Granada concept differed in featuring an hotel and function centre and by being based on the creation rather than the adaptation of a site. The principal site selection criteria were:

1) a land area of approximately 64 acres (160 ha);
2) a site within two hours driving distance of London;
3) a good local road network, giving access to major roads;
4) the ability to create an attractive internal environment with natural features and lots of water;
5) a site where the planning authority would be in favour of development; and
6) a site that could be purchased.

(Cobham Resource Consultants 1988)

Initially the Cotswold Water Park was identified as a suitable location, although the company also considered other areas. Following meetings with the county and district planning authorities in a number of areas, several possible sites within the Water Park were identified, including the site at Somerford Keynes, put forward by Gloucestershire County Council. The shortlist of sites was then examined, with some sites being ruled out because of probable planning problems, such as SSSI designation or unacceptable impacts on existing settlements or other areas. Having taken all these factors into consideration, the Somerford Keynes site was selected as being 'the best potential development site in central southern England' (Cobham Resource Consultants 1988, 5). Following an approach to ARC Properties Ltd, current owners of the site, an agreement was made for Granada to buy the site once planning permission had been obtained. At this point, therefore, it appeared that all of Granada's six criteria had been met and that the process of planning the development and obtaining the necessary permissions could commence.

The next stage of the development process was a detailed examination of the statutory planning documents to ensure that the proposed development would be in accordance with the plans. After consulting both national and local policies and plans, Granada was able to conclude that:

> ... a holiday village development, if properly planned, designed and managed, would be compatible with the policy framework. (Cobham Resource Consultants 1988, 8).

This left the way clear for the preparation of an outline planning application to the Cotswold District Council for the holiday village, to comprise 600 lodges, a 70-suite hotel and function centre and a large indoor leisure centre, all within a self-contained landscaped park. When fully operational, 18 months after construction had started, the village would cater for up to 2800 visitors at a time, with a staff approaching 400. In addition to the outline planning application, Cotswold District Council asked the company to prepare an environmental statement, as laid out in Schedule 3 of the Town and Country Planning (Assessment of Environmental Effects) Regulations 1988. The planning application and environmental statement were submitted in November 1988, with a target date of September 1989 for the development to commence.

In agreeing to prepare the environmental statement, Granada was aware that the scale and location of the development might lead to disquiet amongst local residents. However, the company made the point that the financial viability of the project depended upon its scale, meaning that it was important for local people to be fully aware of the development before it commenced. However, unlike the site at Thorpe, the residents were not faced with a choice between leisure and tipping, although failure to gain a permission for the holiday village might leave ARC Properties with no option but to begin filling.

In agreeing to prepare the environmental statement, however, Granada did have to go a lot further in finalising its plans than might have been desirable, especially when local opposition to some of the minor proposals, such as an amphitheatre, could have undermined the whole application. In choosing to prepare such a detailed statement as well, Granada also laid its plans open to detailed scrutiny, resulting in some significant and informed objections to the scheme on points of detail that simply would not have been available in Leisure Sports Ltd's Thorpe application nearly twenty years previously.

Not surprisingly, the environmental statement, prepared by Cobham Resource Consultants for Granada, presented a favourable impression of the proposed development. Apart from some physical disturbance to the site and an increase in local traffic during construction, most of the environmental impacts were found to be positive. In particular, the environmental statement brought attention to the employment opportunities offered by the proposal, with 750 construction jobs in addition to the 400 permanent jobs created once the village was operational, leading to a 2.5 per cent reduction in the local rate of unemployment. Other benefits including a major environmental improvement to the main road through the Cotswold Water Park, agreed by all to be in need of such an improvement, the establishment and future management of a new nature reserve close to the holiday village and improved access to recreational facilities for local people. In a recent report on the economic impact of the village (Henley Centre 1991) it was estimated that it would generate an annual disposable income of £6.5m, accruing to nearly 8000 people, the majority of whom would not be directly employed at the village. This level of economic impact would compare favourably with a major industrial development, although the environmental impact of the village should be considerably lower.

However, Cotswold District Council had some major reservations about the proposal, mostly stemming from a concern over any environmental deterioration that might be caused by the new use. In particular, the Council was concerned about the effect that the development would have on the residents of Somerford Keynes, the local water quality, levels and supply, the disposal of sewage, the effect of noise and air pollution, the effect of traffic on local roads, the effect on conservation and habitat and the ability of the local fire, police and ambulance services to deal with any emergencies at the holiday village. Consultation with the public also raised a number of other issues, particularly concerning noise and light, use of the hotel as a conference venue and visual screening of the site from the existing communities.

In response to these concerns, and in the hope of securing an early planning permission, Granada established a working party to oversee the revised application. It included members of the local authorities and the various consultants retained by the company. Granada also agreed to furnish the planning authority with additional information relevant to the reservations expressed about the outline planning application. In so doing, however, it made it clear that it felt that the original environmental statement was an appropriate response to the 1988 Regulations and that any further information was provided 'in the interests of ensuring that the local planning authority, statutory consultees and the public are fully informed about the development proposals' (Cobham Resource Consultants 1989, 1).

The revisions to the environmental statement were submitted in January 1990. For the most part, the additional information represented a response to the reservations expressed to the working party. Apart from the changes to the site itself, including revised access to discourage visitors from using minor local roads, most of the revisions were external to the site, including a revised routeing of the water, electricity and sewage services to avoid existing settlements. Some more work was done on the noise and visual intrusion of the site, with independent consultants gauging that there would be no increase in background noise from the site itself, once construction was complete, and that there would be very little visual intrusion once the vegetative screen had been allowed to reach maturity. There would, however, be increased noise from the extra traffic generated.

There was also concern for the effect of the site on wildlife without, it appeared, any recognition of what else might happen to the site if the holiday village did not proceed. However, in order to meet some of the criticisms, Granada agreed to set a substantial area of the site aside as a nature reserve, to employ a warden and to provide the finance for a new reserve south-west of the site. Finally, many of the local residents wanted access to the sports and leisure facilities proposed in the development. Granada therefore offered a membership scheme for local villagers, allowing them to use the pool at subsidised rates and the remainder of the facilities at the same charges as guests.

Regardless of these concessions, the presumption in favour of the development in the statutory plans and the endorsement of the site by Gloucestershire County Council, the application was refused by the planning committee later in 1989. In a most remarkable series of events, however, a subsequent meeting of the full council reversed the earlier decision, but not before the Secretary of State for the Environment had decided to call in the application for review. Since the council had eventually supported the application by granting permission, it was left to local opposition groups to mount an attack on the proposal at the forthcoming public inquiry. The basis of this attack was an audit of Granada's environmental statement. The public inquiry was eventually heard in Cirencester in 1990, two years after the original application, with Spring 1991, the proposed date for opening the village to the public, passing before the Secretary of State's decision had been announced.

Eventually, and with a major reversal of many of the arguments surrounding the proposal, the Secretary of State, in agreeing with the inspector's report, decided that the national and regional economic significance of the village was such that permission should only be withheld if there were any other important interests that might be prejudiced. Since none of the objections was felt to come into this category, outline permission was granted subject to a Section 106 (formerly Section 52) agreement being signed between Lakewoods Ltd (the company formed by Granada and developers John Laing to build the village) and the local authorities, covering details of works to be done before the commencement of construction.

9.4 Guildford Leisure Centre

Guildford was one of the first towns in England to develop a municipal sports and leisure complex, when the current facility in the city centre was opened at the start of the 1970s. Although advanced for its day, and with adequate parking for users, it became obvious by the mid-1980s that it was lacking many of the elements demanded of modern centres, such as a leisure pool, ice rink, bowling alley or adequate catering and retail space. Faced with the prospect of requiring these facilities in order to meet residents' demands, the Borough Council's Arts and Recreation Committee decided in 1986 that the best policy was to relocate the central sports and leisure complex to a more peripheral site with better vehicle access, adequate parking and the space to incorporate all the new activities within one centre.

After an extensive feasibility study, it was decided in May 1988 that a major new leisure complex, comprising squash courts, leisure and competition pools, flumes and water activities, an ice rink, health and fitness suite and a general sports hall, should be built between the A3 and A25 roads at Stoke Park, an area of informal recreation already owned by the council (see Figure 9.4). The area selected was felt to have many advantages, including being sufficiently close to the town centre to attract potential users, being large enough (at some 10 hectares) to accommodate a wide range of facilities and associated car parking and having excellent vehicular access to the A3 and A25 roads without causing increased traffic congestion. The likely catchment area for this proposal was felt to be in the order of two-and-a-half million people, living within a driving time of 40 minutes.

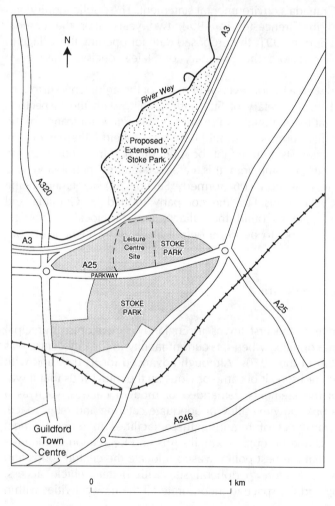

Figure 9.4   The Location of Guildford Leisure Centre

Although proposed by the local authority, the leisure centre development was still subject to the planning system in the same way as it would have been if proposed by a private developer. Before seeking a builder, or even drawing up a detailed design, therefore, the Arts and Recreation Committee submitted an outline application, under Regulation 4(5) of the Town and Country Planning General Regulations 1976, to the Planning Department in mid-1988. Following a favourable officer's report, outline permission was granted in March 1989 subject to approval of Reserved Matters.

It was at this point that the Borough Council sought proposals for the detailed design and construction of the complex, eventually considering four schemes, from Sunley Projects, Norwest Holst, Pearce and Monk. However, given the tight specification drawn up by the Council, both in terms of the type of facility required and its location, there was little to choose between the proposals, even after an environmental assessment of the schemes had been undertaken:

> An environmental comparison does not produce a clear winner. The differences between sites are small in relation to the overall impact of any of the developments. Most of the comparisons could be substantially altered by relatively small design changes or mitigation measures. ... If anything, we prefer those schemes having site access close against the woodland (Norwest Holst, Monk), having good earth screening to south and east (Norwest Holst, Sunley, Pearce), which take least land reinforcement and tree planting (Sunley, followed by Norwest Holst and Pearce).

> We do not consider that the environmental comparison is likely to be a major scheme selection issue. Thus it is unlikely that a choice made on cost or operability grounds will be contradicted by environmental drawbacks. (Montagu Evans 1990, para 3.6).

Using other parameters such as cost, the Borough Council subsequently selected the proposal by Sunley Projects Ltd. Subsequently, the local planning authority requested that a full environmental statement was submitted, recognising that although not required under the 1988 Regulations the proposal was of such a scale that an environmental assessment was appropriate. The statement was later published for information purposes (Montagu Evans 1990A). Although the environmental report identified the loss of open space as a major impact of the scheme, along with increased traffic noise, ecological damage to trees and wildlife and the visual, traffic and construction impacts of the development, detailed permission was eventually forthcoming in mid-1990, some 26 months after the initial application.

Apart from the tender process, the major reason for the length of time between application and approval for what was, after all, the provision of a range of facilities demanded by the local populace concerned, in the light of the environmental report, the appropriateness of the scheme in terms of its size, location and siting. Objections were raised by local people that the proposal centralised all major attractions, inevitably leading to the closure of some town centre facilities. Equally, objections were raised about the proposed location away from the town centre, with its implications for accessibility for those without private transport. Finally, there was much resentment over the actual choice of site since it entailed the loss of existing public open space. The environmental consultants did point out, however, that since the majority of the proposed site had been allotments there was little actual loss of public access (Montagu Evans 1990), whilst plans had already been approved by the Borough Council to extend Stoke Park by some 227 acres (90 hectares) alongside the River Wey.

The strength of resentment led to the formation of a pressure group 'Stoke Park for the Residents of Guildford' (SPROG) which tried, alongside other groups such as the Surrey Branch of the Council for the Protection of Rural England to have the single proposed development replaced by a range of facilities spread across the urban area.

Of equal concern with these objections was the relationship between the proposal and the district-wide Guildford Borough Local Plan, adopted in 1987. Although covered by a general policy to provide and improve leisure facilities in Guildford (policy R3), there were concerns about the extent to which the proposal represented a departure from other, more specific policies. In particular, the commentary to policy R2 suggested that a relocation of facilities from the centre to the edge of town should be resisted, as should any attempt to centralise facilities rather than distribute them around the Borough (policy R1). There was, furthermore, town centre land that had already been identified for the development of an ice rink (policy R8/TC10), whilst the site selected at Stoke Park was to remain as an area for informal recreation (policy R14) and the allotments on the site were to be retained (policy R23).

Despite these concerns, there was neither a refusal of permission nor a call-in by the Secretary of State for the Environment. Indeed, the planning department suggested that although the proposed use of the site was more formal and intensive than had been envisaged in the local plan, it was not necessarily contrary to the spirit of the policy. Equally, the planning department felt that the identification of a site for an ice rink was more indicative of the local need for such a facility than the intended location, given the ease of access to the proposed site. Finally, in recognising the potential loss of allotments the planning department suggested (rather than required) that, although they were non-statutory as well as being currently under-used, steps be taken to find a suitable replacement site.

Regardless of the information and evidence to the contrary, therefore, the planning department approved the scheme with only minor alterations and restrictions. It is certainly clear from this that local need, as expressed by the local authority, is a powerful force in gaining consent for a proposal that on the face of it appears to depart substantially from a current development plan. What is less certain is the extent to which economic or financial considerations were important. Although initially discussed before the advent of compulsory competitive tendering, there is no doubt that an integrated facility of the type being developed would be of far more interest to both the direct labour organisation (which subsequently won the tender process) and potential commercial operators than a dispersed range of specialist facilities.

As this example amply illustrates, proposals of this type provide many difficulties for local authorities. Undoubtedly, the new leisure complex is likely to give Guildford residents access to one of the biggest and best leisure centres in the south of England, without the need to travel any more than a short distance. In the process, however, the 'local' feel of the old sports centre and other such facilities will probably be lost and there must be a residual feeling on the part of many residents that they have been deprived of the protection that should have been afforded them by the local plan. Whether the decision to build the new leisure centre will be justified in the course of time remains to be seen, much now depending upon how the facility is to be managed and how use by local residents is protected and encouraged.

Now under construction, the leisure centre will feature: leisure and competitive swimming pools; a diving pool; a sports hall; ice rink; squash courts; a 32-lane bowling alley; health and fitness centre; a running track and all-weather pitch; and catering, retail, nursery and ancillary office space. Construction is being undertaken by Sunley Projects Ltd on a design and build contract (see Chapter 4) worth an estimated £28m, with completion due in autumn 1992. Having recently been successful with its competitive tender (see Chapter 7), the Borough Council will assume managerial control upon completion, when the centre will become known by its recently selected name of Spectrum.

9.5 Royalty and Empire

The genesis of the Royalty and Empire exhibition at Windsor occurred in the Nineteenth century when both the Great Western and the South Western railway companies managed to extend their rail networks to Windsor, so creating an eventual over-capacity of trains, lines and stations (South 1978). Although both stations survived transfer to British Rail, as Windsor and Eton Central Station (formerly the Great Western station) and Windsor Riverside (formerly the South Western station) (see Figure 9.5) respectively, the British Rail Property Board recognised that their future was far from certain, although their historical and architectural heritage was such that they could not be abandoned altogether.

By the end of the 1970s the decision had been taken that British Rail could not continue to support the two stations at Windsor in their existing form. Given their respective locations, the central station appeared to be the most suitable for retention, although its location equally offered scope for conversion to another use. In favour of retention was the fact that the station

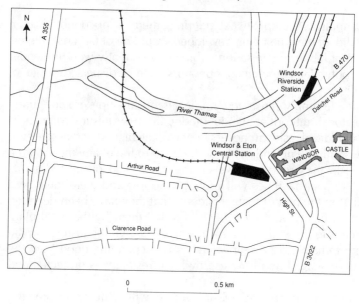

Figure 9.5   The Location of Stations in Windsor

itself was listed, thereby limiting conversion potential, although its status as a branch line to Slough was considerably less than the direct connection to London Waterloo offered by the Riverside station. In the event, British Rail elected to retain the whole of the Riverside station and, because of the construction and scale of the Central station, to retain a single line into the centre of Windsor whilst putting the majority of the station to an alternative use.

As with similar schemes in other Brunel stations, these large open plan buildings were ideally suited to retail use. Indeed, J. Sainsbury had already undertaken the successful conversion of a similar Great Western station in Bath and were keen to do the same in Windsor. However, whilst recognising that conversion of part of the central station to a retail outlet would both protect the fabric of the listed building as well as provide a major town centre shop, the local authority was not in favour of more retail development in the centre of Windsor. In reconsidering the conversion of the station, therefore, the British Rail Property Board entered into discussions with Madame Tussauds about the possibility of retaining the station, but developing a themed exhibition depicting the railway era in Windsor. In addition to retaining the station in a fairly original form, both Madame Tussauds and the British Rail Property Board felt that such a scheme would accord with Windsor's tourism industry, then based on Windsor Castle and the nearby safari park.

Although recognising the attractions of the proposed scheme for the retention of the building, the local authority was much less enthusiastic about attracting extra visitors to Windsor, on the basis that the town was already subject to severe congestion at the height of the tourist season. However, the local authority appeared to be alone in this view, with traders and tour operators keen to see additional facilities in Windsor, especially on days when the Castle was closed to the public. In addition, Madame Tussauds argued that just as its main venue in London did no more than attract people already in London, so its Windsor venture would do the same, thereby not attracting many additional tourists but keeping them in Windsor for longer, so encouraging them to put more money in the pockets of Windsor's traders.

Eventually, recognition that a major additional attraction in Windsor which could retain the spirit and feel of the central station, when developed by a company with the reputation of Madame Tussauds, was preferable to the construction of another supermarket led the local authority to withdrawing its objections and allowing the project to proceed. The approval was, however, on the strict basis that amongst other restrictions, exterior alterations to the station would not be allowed, whilst the lease specified that the building should always contain an exhibition associated with Royalty. Despite these restrictions, Madame Tussauds and the British Rail Property Board agreed a 55 year lease of part of the station, starting in 1979, on the payment of a fixed annual base rent plus a percentage of net operating profit. The remainder of the station, comprising one platform and a ticket office, was to be retained by British Rail as a working station.

Originally opened in April 1983 as Royalty and Railways, the exhibition depicted the arrival of guests to celebrate the Diamond Jubilee of Queen Victoria in June 1897. Having paid the entry fee at a replica ticket office, visitors followed a route around the station with different scenes depicting the Royal waiting room, the arrival of the train, the greeting of the guests and the transfer by horse and coach to the Castle, the last scene being witnessed from a balcony above the station. Featured in all these scenes were, of course, examples of Madame Tussauds wax models, together with a reconstruction of the station as it would have appeared in 1897.

At the end of the route around the station visitors arrived at a newly constructed theatre to see a multi-slide presentation on the reign of Queen Victoria, followed by scenes from her life depicted by moving waxwork models; a major technical innovation at that time and a highlight upon which to end the visit. Because of the lack of space there was no room for a catering facility, whilst the shop was extremely cramped.

Following a disappoin ing opening and first season with poor visitor numbers, Madame Tussauds renamed the exhibition as Royalty and Empire in an attempt to play down the apparently negative association with the railways and to attract the attention of more of the foreign visitors. In addition, the local authority was persuaded to modify its views on external alterations and grant permission for a new enlarged entrance to attract people the short distance from the High Street. These modifications appeared to be successful, allowing Madame Tussauds to report a 20 per cent rise in visitors in 1985, as well as improved sales in the extended and relocated shop.

Visitor numbers continued to rise in the next few years, peaking in 1987. However, the numbers never reached Madame Tussauds' reported target of attracting one-sixth of Windsor's three million annual visitors. Furthermore, the venture remained small compared to the growing list of other Madame Tussauds exhibitions and ventures, such as Chessington World of Adventures, Warwick Castle and Wookey Hole Caves in Somerset. In retrospect this apparent ceiling to visitor numbers was probably inevitable, as a consequence of the nature of the exhibition and the physical layout of the building. With the requirement for all visitors to follow the same route around the exhibition and, eventually, to fit into a theatre capable of taking a maximum throughput of 750 people per hour, the venture was realistically limited to a maximum of 4000 people per day in high season, with many being forced to wait a long time before admission to the theatre. This maximum number was achieved just once. For a venture with such high capital and operating costs this level of throughput needed to be supplemented by other forms of income. In part this came about through diversifying into the functions market in the evenings. However, the ability to exploit this market was limited by the lack of catering space, as well as a lack of suitable accommodation and parking close by.

The extent to which Royalty and Empire's problems were caused or exacerbated by the local authority is a matter for conjecture. However, by relying on the established pattern of tourist visits to Windsor, involving short stays by predominantly foreign visitors to see the Castle and by others, primarily local, to take trips on the pleasure boats on the River Thames or visit the safari park, the local authority certainly did not appear to take account of what might happen in the future. When international terrorist activity and, more recently, the Gulf War depleted the number of foreign tourists to Britain, Windsor fared particularly badly since it had failed to develop other markets or to encourage new types of venture to spread the risk of this occurrence.

By the start of the 1990s it was noticeable that Madame Tussauds' policy of acquisition and operation was changing. With the £60m purchase of Alton Towers from the beleaguered John Broome and the sale of the Wookey Hole enterprise, Madame Tussauds were both moving into large scale recreation and tourist facilities as well as acquiring the freehold interests in their sites. On neither account did Royalty and Empire match these criteria. Faced in addition with the local climate of control and introspection in Windsor, Madame Tussauds began to question the wisdom of developing the Royalty and Empire venture and, in early 1991, decided that a sale to L & R Leisure was the best future course of action. Since that time some alterations have taken place, effectively altering the venture from an exhibition with associated retail, to a mixed retail and exhibition outlet. Indeed, the summer of 1991 saw stallholders renting space from L & R in Station Approach, whilst it is reported that the name of the exhibition is to change to the Royal Station Shop. In the space of just 12 years, therefore, the property which might originally have been converted into retail premises has seen a major investment in leisure fail to achieve the desired returns and is now to be used mainly for retail purposes.

Quite what the future holds for the site remains to be seen, although the ability of the new owners to make major alterations is limited both by the nature of the building, the restrictions in the lease and the local authority. However, with Windsor now losing tourists the local authority may be more favourably disposed to innovation and alteration as a means of attracting new markets, whilst it appears that both the building and the lease are flexible enough to provide scope for change.

9.6 Conclusions

It is tempting to conclude from the case studies that little has changed over the twenty year period from the original application to develop Thorpe Park to the present day; that there remains a strong presumption against development in the countryside, even where the potential for significant benefits can be shown to exist, whilst there is a presumption in favour of development in urban areas, even if the location or intended use of the site is contentious. Indeed, preservation of the countryside has remained a central feature of local planning policies, with local authorities remaining wary of development proposals even where they are supported by central government policy and exhortation:

> Whatever the plans and proposals of central government, the overwhelming issues dominating local planning and policy making are likely to remain linked to the preservation of the existing environment. (Byrne and Ravenscroft 1989, 24).

What has become clear from all of the proposals and developments is the central role played by local inhabitants, either directly through public pressure, or indirectly through a reluctance on the part of the planning committee to sanction any development that might harm them. The Guildford case is a slight exception to this, although as a public investment it can be argued that it has been developed with local interests in mind, even if some of the protesters did not feel that this was the case.

In essence, therefore, the problems over gaining consents to develop large-scale leisure and tourism facilities must lie in the very interaction between development initiatives and the fabric of the planning system itself. Whilst both national and local planning policies may be in favour of the concept of large-scale leisure and tourism development, the actual planning committee that is asked to give the necessary permission remains wary. Although the reasons for this unease have yet to be fully elucidated, a large part of the problem may well be the relative uniqueness of every individual proposal. If the committee is faced with a proposal for an office block, a factory or even a large housing estate, evidence already exists of the impacts that are likely to be felt by the existing community. Indeed, the developer is probably in the position to offer a track record of similar developments as evidence of the impacts, as well as good practice in construction and disposal. Equally, these types of development tend to fit squarely within the formal planning process; areas are zoned for industrial and commercial developments, whilst houses, schools and shops are needed for the resultant increase in the labour force.

Large-scale leisure developments do not exhibit any of these characteristics. By their very nature, most leisure developments are marketed as being a new or different concept, even if in reality they are more realistically derivations of existing concepts. Consequently, there is unlikely to be a suitable zoned area for such developments, with any leisure-related zoning more likely to involve public sports facilities or particular enterprises which the local planning authority has identified as absent or under-represented in the area, such as the ice rink identified in the Guildford local plan. Thus, whilst the developer may be able to provide a track record of good practice or offer to retain the management of the completed development (as in the case of Lakewoods Ltd at Somerford Keynes), the planning authority is unlikely to have the safety of precedent or the rationale of local need to justify approval of an application.

Despite these apparent difficulties, work recently undertaken for the Leisure and Tourism Industries Sector Group of the National Economic Development Office (NEDO) indicates that the passage through the planning system of major leisure and tourism development proposals in the countryside is very little different from other types of large-scale proposal, such as industrial development (Davies *et al.* 1991). Indeed, 13 of the 18 selected case studies received planning permission, although some involved substantial development in green belts whilst others involved the potential loss of considerable portions of Sites of Special Scientific Interest. In concluding the report, therefore, Davies *et al.* (1991) commented that not only are large-scale leisure and tourism developments receiving planning permissions, but they are receiving them in areas of designated countryside that were hitherto assumed to be inviolable.

Whilst this might be the case, however, many local planning authorities have rejected applications, on the grounds of public resistance if nothing else, and put the onus on either the developer to challenge their judgement or the Secretary of State to call in the application. Whilst the appeals and call in procedures provide a formal opportunity for all interested parties to have their say, it is questionable what extra information can be made available at inquiry, given the comprehensive nature of the requirements for the environmental statements originally presented to the local planning authority. This did not, however, stop the public inquiry into the Somerford Keynes proposal lasting nine weeks.

What is even more frustrating for the developer, as well as being further indicative of the problems inherent in the system, is that the decision by the Secretary of State was not forthcoming for over a year after the public inquiry. Whilst the delays may have seemed incomprehensible after the preparation of a comprehensive environmental statement and an apparently exhaustive inquiry process, the fact remains that the inspector and, in turn, the Secretary of State, do not have any greater knowledge or understanding of the impact of large-scale leisure and tourism developments than the local councillors. Equally, however, the Secretary of State has to take a wider view of the issues than was, perhaps, possible for the local authorities, particularly with respect to national and regional issues. This is amply demonstrated in the Secretary of State's decision letter following the Somerford Keynes inquiry, where considerable emphasis was placed on the national significance of the proposal, to the point where all local objections could be overruled:

> Support for this style of development in general, and this proposal in particular is given by the English Tourist Board which is a statutory body. In view of their status I consider the strategy they operate to be on a par with national policy, if indeed it is not national policy. They claim that the proposal is in the national interest, being fully in accordance with their current strategy. I see this degree of support by such a body to be tantamount to official recognition of the need for the proposal. In view of this, and recognising the demand for this type of development, I consider the proposal to be necessary or even essential in the context of the relevant Development Plan policies if the tourism objectives are to be achieved (Inspector's Report para. 712, p. 156).

and:

> The Secretary of State agrees with the Inspector's conclusions and accepts his recommendation. He takes the view that, in general, development proposals should be allowed to proceed, having regard to all the material considerations, unless they would conflict with an interest of acknowledged importance (Decision letter, para. 7). ... The Secretary of State is satisfied that no interest of acknowledged importance would be harmed by the proposed development ... (Decision letter para. 14).

What is equally apparent is that the level of mistrust between developers, operators and the local authority does not appear to diminish once a venture is in operation. Whilst relations seem to have improved between Thorpe Park and the local authority, there must be questions as to whether a developer with fewer resources than the RMC Group would have persevered. Equally, the attitude of the local authority in Windsor could have prevented Madame Tussauds from developing Royalty and Empire at all and may have ultimately contributed to the decision to sell. Quite why these situations should occur is not clear, although the constant need to change and up-date recreation ventures may leave the local authority feeling that the current and proposed future use of sites bears little relationship to the uses for which permission was originally granted.

What is equally clear is that in many cases plans and policies derived in consultation with local people, to meet local needs, can be ignored in the interests of national policy or political expediency. This was clearly felt to be an important factor in the Somerford Keynes case, when Max Pearce, managing director of Lakewoods Ltd stated that:

> In the last three years we've seen attitudes change from government offering inducements to Center Parcs to go to Sherwood Forest, to people becoming more organised in their resistance to development in the countryside. (Januarius 1991, 25).

There is a strong view that the Planning and Compensation Act 1991 will put a halt to the worst of these excesses by implementing a more plan-led system of development control. The extent to which this can become reality is largely dependent on the speed with which suitable plans can be drawn up and agreed; evidence from the Town and Country Planning Act 1971 suggests that this could be a long time in the future.

Faced with this situation it is apparent that a solution has to be found soon in order to avoid the current levels of misunderstanding between developers and local planning authorities. In the recent report on the rationale and economic impact of holiday village schemes (Henley Centre 1991), the authors warn that Britain stands to lose out in the competition to provide such holiday facilities unless planners and developers can work in partnership to ensure that new facilities meet local criteria, so avoiding costly planning delays. However, whilst this may be a laudable aim, evidence from the case studies in this chapter, as well as in many other cases, serves to indicate that even where there appears to be harmony between planners and developers, long delays continue to occur. Indeed, probably the single biggest indictment of the system must be that, were Thorpe Park to be proposed today, it is likely that the planning delays would be just as long as in the early 1970s, with neither local councillors nor planning inspectors in any better position to judge the merits of the scheme against its associated costs.

However, the enactment of the Planning and Compensation Act 1991 provides an excellent opportunity to redress many of these issues in the future. By attempting to implement a plan-led system of development control, the Department of the Environment has given both the leisure and tourism industry and local planning authorities the chance to agree on appropriate types of development and suitable locations in advance of any actual applications, so hopefully preventing some of the lengthy time delays that have accompanied many recent development proposals. However, this will only occur if the industry seizes the initiative and ensures that it plays a central part in future planning policy generation.

10 Conclusion: Recreation Planning and the Development Process

Recreation provision is something of a paradox. The very notion of 'planning' for a concept that is essentially personal to each of us seems misplaced or, at the very least, doomed to failure. Yet recreation, or at least leisure, is central to our lives, as either the antithesis or partner of work. Even so, we do not appear to possess anything as strong as a psychological 'need' for recreation, although to be fit and healthy most of us have a physiological need for exercise. Yet one of the original cornerstones of the planning system was a wish to protect existing urban open spaces, presumably predicated on an assumed psychological, rather than physiological, need for recreation. Despite its origins, however, recreation has not maintained its place in the planning system, or even in the wider sphere of public provision. This suggests that, rather than being central to our lives, it is really somewhat peripheral, a view supported by the lack of a model of the development process suitable for recreation provision. Yet both the public and private sectors of the economy have developed a wide range of facilities, to the point where recreation provision is one of the major local government cost centres and commercial recreation and tourism provision represents one of the fastest growing sectors of the economy.

Accepting that it is both possible and desirable to plan for recreation, by planning provision for recreation opportunities rather than the planned provision of recreation itself, the paradox can be refined by the recognition that recreation provision has strong elements of both welfare and the market (Coalter *et al.* 1986), rendering generalisations difficult to make and many observations apparently contradictory. As far as the welfare element of recreation provision is concerned, the public sector has, for most of this century, provided a wide range of facilities on the basis of ill-defined and, occasionally, erroneous objectives. It has also contributed to some market-orientated facilities by grant-aiding the education and wardening services of some commercial operators, as well as seeking access agreements and other means of allowing people into the countryside. With the exception of the Victorian seaside holiday trade, commercial provision is a much more recent phenomenon, arising principally from the increasing affluence and mobility of post-war Britain. Whilst guided, encouraged and constrained by government planning, grant-aid and other policies, the basis of this commercial provision has been the private ownership and control of land; the property power (Denman 1978).

The public sector involvement in recreation is complicated by its dual role as planning authority and direct provider of consumption services; a mixture of planning as provision and planning as the guidance and control of land uses. The evolution of both these roles has been largely incremental, with little apparent coherence to either (Blackie *et al.* 1979). The responsibility for planning land uses has therefore occupied an uncomfortable position between the various tiers of government, with advice, guidance and changes of emphasis emanating from Westminster with sufficient regularity to ensure little continuity in the preparation of development plans and the application of development control. Similarly, the basis of direct provision has been complicated by the multiplicity of government departments involved as well as the changing political fortunes and requirements of successive governments.

However, underlying both these roles have been some threads of a consistent policy. Perhaps the most basic of these has been the use of both planning and recreation to maintain the *status quo*. This has consisted mainly of ensuring the continued sovereignty of land ownership, particularly in the countryside, as well as the prosperity of the property industry. Equally, it has involved using both planning and recreation provision, often in tandem, to achieve particular aspects of national policy. Thus the popularity of recreation provision has rested on such qualities as its role in public health and morality, its use in leadership training within schools, its value as a tool of economic regeneration and its ability to distract disaffected youth and the unemployed; the popularity of planning has been based upon many of the same attributes, including public health, economic regeneration and the improvement of urban living.

What makes planning and recreation development even more similar, however, is the recognition that both of them have been, and continue to be, used by government as 'treatments' for social ills or injustices (Long and Hecock 1984). Until the election of the avowedly *laissez-faire* Conservative government in 1979, it might have been possible to believe that the application of the statutory planning system had been based on a collectivist belief that state intervention in the land market could improve the social condition. The retention and, indeed, enhancement of the system since that time, by a government committed to removing any constraints to the operation of the market, must signal that the planning system has other attributes that the government does not wish to forego; that it is a method of imposing government policy rather than provoking social change. The same critique can be applied to public recreation provision, in that although supposedly based on the standard economic arguments for state intervention in the market there has, seemingly, been no conception of such provision ever changing or ameliorating anything other than immediate and identifiable ills (Henry 1984). The recent Planning and Compensation Act 1991 appears to add weight to this view by attempting to impose a much more plan-led system of development control. Early commentary on the Act (Davies *et al.* 1991) suggests that national strategic issues such as economic development, employment and the balance of payments will have even more influence on future planning decisions, especially for new service industries such as recreation and tourism.

The reasons for commercial provision are much more straightforward, at least in as far as they concern the need for financial survival. However, whilst there has been a constant flow of commercial developments over the last century, the public, wildlife and landscape pressure created by those developments has rarely been intense. Apart from consistent development pressure in the Green Belt around some of the major cities, as well as some pressure in sensitive areas such as the National Parks, the number of proposals for commercial recreation development has been limited. Whilst the full list of reasons for this situation may be long and complex, one of the primary limiting factors has been the activity of the public sector. Although few commercial operators would have been interested in public parks, swimming pools or small sports halls, certainly without subsidy, the public provision of squash courts, golf courses and, latterly, fitness centres, all at subsidised prices, has left limited scope for the private sector in urban areas. Similarly, the provision of country parks and the negotiation of access agreements in the countryside has reduced the scope for the commercial sector in rural areas.

However, with the return to the values of the market heralded and espoused by the Thatcher government, with the consequent deconstruction of the welfare state and the machinery of local government, the reliance on the public sector to plan and provide recreation services can no longer be taken for granted. Whilst this may, or may not, be a good thing, what it has done is bring forward the stark realisation that there is really no effective means of planning recreation provision other than by direct government provision. Effectively we are now faced with the prospect of a shift from 'planning' and 'provision' being synonymous to one where they are distinct, both in terms of function and operation. It is probably for the first time ever, therefore, that there has actually been a genuine need for comprehensive and objective planning policies to guide, encourage and control what is increasingly likely to be private sector recreation development.

There is some evidence to suggest that central government is alive to this need, since recreation, leisure and tourism have recently emerged from over a decade of obscurity to become, once again, important elements of future development plans. Quite what this will actually mean in practice is, as yet, difficult to determine. The one indication received to date, namely Planning Policy Guidance Note PPG 17 on Sport and Recreation, does not augur well, with few new initiatives and very little in the way of incentives for development. Equally, the continuing delay in planning decisions for large-scale leisure facilities, as witnessed in Chapter 9, suggests that there is much work yet to be done in preparing a planning system that is able to cope with the new and increasing demands of the market.

As far as the market itself is concerned, two principal areas of concern remain to be addressed: work on establishing the credentials and acceptability of recreation, leisure and tourism as a serious, long-term industry; and, as a corollary, the derivation of a model of the development process that is suitable for recreation and leisure property. Time and familiarity will, to some extent, help in achieving the first of these requirements. However, work is still necessary to assess the likely social and economic impacts of major recreation developments and upon whom those impacts are likely to impinge. This will inevitably involve attempts to develop and refine techniques such as cost benefit analysis, whilst more detailed attention to the preparation and content of environmental statements may yield many future benefits. A start has recently been made on this, initiated by the Leisure and Tourism Industries Sector Group of the National Economic Development Office. Equally, as more work is done, and as more recreation developments are completed, the nature of an appropriate model of the development process should become more apparent. This, in turn, should be instrumental in aiding the recreation and tourism industry to gain greater acceptance, particularly with institutional investors and other lenders of capital.

What is very clear from this is that a comprehensive approach to planning for recreation development is going to be vitally necessary if the public sector is to establish and maintain its role of guidance and control. What is rather less clear is quite what the underlying planning policies should be and what role, if any, will remain for the public sector in terms of direct provision. Whilst apparently separate, these two questions have much in common, for the only reason to guide national planning policy in terms any different from other industries, or to remain involved in direct provision, must be on the explicit grounds of market imperfections. That these exist, even for the most market-oriented of governments, must be without question. What has become apparent, however, is that recreation provision has become too heavily associated with other areas of government policy for the possible benefits to be clear:

> ... if sport is promoted and pursued as an end in itself it may bring social benefits which will elude the grasp of policy makers if they treat it as little more than a clinical, social or political instrument to fashion those very benefits. (McIntosh and Charlton 1985, 193).

There is a strong future for the public sector in recreation planning and provision, but one that must be defined in terms of viewing recreation for what it is and what it can achieve. This means an emphasis in direct provision on health, physical activity and enjoyment, with appropriate facilities and associated support. For those public facilities now contracted out to the private sector, the emphasis must be on improving access to them for all people whilst, finally, planning policies to guide and control future commercial recreation development should be similar to those applied to any industry, in terms of representing a balance between national and local interests.

References

ADDISON, C. 1931. *Report of the National Park Committee.* Cmnd. 3851. London. HMSO.

ADIE, D. 1985. A bigger splash. *Architect's Journal* 181(11): 64–71.

AKEHURST, R.L. and BLACKBURN, K. 1979. Geographic cost variations in the North Western Regional Health Authority. *Hospital and Health Services Review* 75(11): 400–405.

ALLEN, L.R. and BEATTIE, R.J. 1984. The role of leisure as an indicator of overall satisfaction with community life. *Journal of Leisure Research* 16(2): 99–109.

ALLISON, L. 1975. *Environmental planning. A political and philosophical analysis.* London. George Allen and Unwin Ltd. 134pp.

ANON. 1980. Thorpe Park: Britain's first 'theme park'. *Water Space* 14: 17–21.

ANON. 1989. Farm diversification and planning control. *Estates Gazette* 8937: 117.

ARCHBISHOPS' COMMISSION ON RURAL AREAS. 1990. *Faith in the countryside.* Worthing, W. Sussex. Churchman Publishing Ltd. 400pp.

ARNOLD, S. 1985. The dilemma of meaning. pp 5–22 in Goodale, T.L. and Witt, P.A. (editors). 1985. *Recreation and leisure: issues in an era of change.* Revised edition. State College, Pennsylvania. Venture Publishing, Inc. 423pp.

ARTHUR YOUNG McCLELLAND MOORES AND CO. 1984. *Review of the economic efficiency of national park authorities.* Publication CCP 160. Cheltenham. Countryside Commission. 56pp.

AUDIT COMMISSION. 1983. *Improving economy, efficiency and effectiveness in local government in England and Wales. Volume 1.* London. The Audit Commission for Local Authorities in England and Wales.

AUDIT COMMISSION. 1984. *Improving economy, efficiency and effectiveness in local government in England and Wales: Audit Commission handbook, volume 2.* London. The Audit Commission for Local Authorities in England and Wales. 76pp.

AUDIT COMMISSION. 1987. *Competitiveness and contracting out of local authorities' services.* Occasional Papers No. 3. London. HMSO. 7pp.

AUDIT COMMISSION. 1990. *Local authority support for sport. A management handbook.* London. HMSO. 80pp.

BACON, A.W. 1980. *Social planning, research and the provision of leisure services.* Manchester. The Centre for Leisure Studies, University of Salford. 25pp.

BALE, J. 1989. *Sports geography.* London. E. and F.N. Spon. 268pp.

BANNOCK, G., BAXTER, R.E. and REES, R. 1978. *The Penguin dictionary of economics,* 2nd. edition. Harmondsworth, Middx. Penguin Books Ltd. 467pp.

BARCLAY, I. 1990. The continuing role of the local authority in the leisure market. Paper presented at the IBC Conference: *Leisure property into the 1990's.* 24th. January 1990, Inn on the Park, London.

BARING, N. 1990. 1989 finances. *The National Trust Magazine* 60: 32–34.

BARING, N. 1991. 1990 finances. *The National Trust Magazine* 63: 31–33.

BARLOW REPORT. 1940. *Report of the Royal Commission on the distribution of the industrial population.* Cmnd. 6153. London. HMSO.

BARRETT, S., STEWART, M. and UNDERWOOD, J. 1978. *The land market and development process. A review of research and policy.* Occasional Paper No. 2. School for Advanced Urban Studies, University of Bristol. 174pp.

BARTELMUS, P. 1986. *Environment and development.* Boston, Mass. Allen and Unwin Inc. 96pp.

BEESTON, D.T. 1983. *Statistical methods for building price data.* London. E. and F.N. Spon Ltd. 175pp.

BEVINS, M.I. 1971. Private recreation enterprise economics. pp 33–39 in Larson, E.vH. 1971. *The forest recreation symposium.* State University of New York College of Forestry, Syracuse, New York. 12–14 Oct. 1971. Washington, D.C. U.S. Government Printing Office. 211pp.

BIRCH, J.G. 1971. *Indoor sports centres.* Sports Council Studies No. 1. Lond. HMSO. 67pp.

BLACKIE, COPPOCK, J.T. and DUFFIELD, B.S. 1979. *The leisure planning proces.* ndon. The Sports Council and the Social Science Research Council. pp.

BLAIR, R.D. and KENNY, L.W. 1982. *Microeconomics for managerial decision making.* New York. McGraw Hill Book Company Inc. 447pp.

BLUNDEN, J. and CURRY, N. 1988. *A future for ou. untryside.* Oxford. Basil Blackwell Ltd. 224pp.

BLUNDEN, J. and CURRY, N. (editors). 1989. *A people's charter?* London. HMSO. 299pp.

BODLENDER, J. 1990. Managing the future. pp 247–257 in Quest, M. (editor). 1990. *Horwath book of tourism.* London. The Macmillan Press Ltd. 264pp.

BONYHADY, T. 1987. *The law of the countryside.* Abingdon. Professional Books Ltd. 290pp.

BOVAIRD, A.G. 1991. Evaluation, performance assessment and objective-led management in public sector leisure services. Paper presented to the Leisure Studies Association annual conference: *Leisure in the 1990's:*

rolling back the Welfare State. University of Ulster at Jordanstown. 12–15 September 1991. 30pp.

BOVAIRD, A.G., TRICKER, M.J. and STOAKES, R. 1984. *Recreation management and pricing.* Aldershot, Hants. Gower Publishing Company Ltd. 182pp.

BRAMHAM, P. and HENRY, I.P. 1985. Political ideology and leisure policy in the United Kingdom. *Leisure Studies* 4(1): 1–19.

BROMWICH, M. 1979. *The economics of capital budgeting.* London. Pitman Publishing Ltd. 395pp.

BUTLER, D. and RICHMOND, D. 1990. *Advanced valuation.* Basingstoke. Macmillan Education Ltd.

BUTT, H.A. and PALMER, D.R. 1985. *Value for money in the public sector. The decision maker's guide.* Oxford. Basil Blackwell Ltd. 187pp.

BYRNE, P.J. and CADMAN, D. 1984. *Risk, uncertainty and decision-making in property development.* London. E. and F.N. Spon Ltd. 184pp.

BYRNE, P.J. and RAVENSCROFT, N. 1989. *The land report. Diversification and alternative land use for the landowner and farmer.* London. Humberts. 29pp.

CADMAN, D. 1984. Property finance in the UK in the post-war period. *Land Development Studies* 1: 61–82.

CADMAN, D. and AUSTIN-CROWE, L. 1983. *Property development,* 2nd. edition. London. E. and F.N. Spon Ltd. 268pp.

CADMAN, D. and CATALANO, A. 1983. *Property development in the U.K. – evolution and change.* Reading. Centre for Advanced Land Use Studies, College of Estate Management. 34pp.

CARD, R., MURDOCH, J. and SCHOFIELD, P. 1986. *Law for estate management students,* 2nd. edition. London. Butterworths. 696pp.

CARRINGTON, B. and LEAMAN, O. 1983. Sport as community politics. A critique of some recent policy initiatives in sport and physical recreation. Paper in Haywood, L. (editor). 1983. *Sport in the community. The next ten years: problems and issues.* Leisure Studies Association Newsletter Supplement.

CARRUTHERS, S.P. (editor). 1986. *Land-use alternatives for UK agriculture.* CAS Report 12. Centre for Agricultural Strategy, University of Reading. 45pp.

CARSON, R. 1962. *Silent Spring.* London. Hamilton.

CASEY, D. 1980. Cost effectiveness in sports and leisure centres. pp 17–27 in Association of Recreation Managers. 1980. *Cost effective recreation management.* National Seminar Conference Report. Over, Cambs. Association of Recreation Managers. 90pp.

CATLIFF, T. 1983. The Thorpe Park experience. *Chartered Surveyor Weekly-*4(10): 489–491.

CHARTERED INSTITUTE OF PUBLIC FINANCE AND ACCOUNTANCY. 1983. *Leisure and recreation statistics 1982–83 estimates.* London. CIPFA Statistical Information Service. 108pp.

CHARTERED INSTITUTE OF PUBLIC FINANCE AND ACCOUNTANCY. 1988. *Leisure and recreation statistics 1988–89 estimates.* London. CIPFA Statistical Information Service. 123pp.

CHARTERED INSTITUTE OF PUBLIC FINANCE AND ACCOUNTANCY. 1990. *Leisure and recreation statistics 1990–91 estimates.* London. CIPFA Statistical Information Service. 119p.

CHASE, G. 1990. The role for leisure in a balanced investment portfolio. Paper presented at the IBC Conference: *Leisure property into the 1990's.* 24th. January 1990. Inn on the Park. London.

CHESSELL, H. 1946. *National parks for Britain.* Birmingham. Cornish Brothers Ltd. 60pp.

CHICK, M. and SCRASE, A. 1990. *Agricultural diversification and the planning system.* Bristol. Department of Town and Country Planning, Bristol Polytechnic. 149pp.

CHUBB, M. and CHUBB, H.R. 1981. *One third of our time? An introduction to recreation behavior and resources.* New York. John Wiley and Sons, Inc. 742pp.

CLARK, G.L. 1982. Rights, property and community. *Economic Geography* 58: 120–138.

CLAWSON, M. and KNETSCH, J.L. 1971. *Economics of outdoor recreation.* Baltimore, Md. The Johns Hopkins University Press. 328pp.

CLAYDEN, P. 1985. *Our common land. The law and history of commons and village greens.* Henley-on-Thames. The Open Spaces Society. 110pp.

COALTER, F., LONG, J. and DUFFIELD, B.S. 1986. *Rationale for public sector investment in leisure.* London. The Sports Council and the Economic and Social Research Council. 181pp.

COBHAM RESOURCE CONSULTANTS. 1988. *A holiday village in the Cotswold Water Park. Environmental Statement.* Abingdon, Oxon. Cobham Resource Consultants. 52pp.

COBHAM RESOURCE CONSULTANTS. 1989. *A holiday village in the Cotswold Water Park. Additional information.* Abingdon, Oxon. Cobham Resource Consultants. 34pp.

COHEN, E. 1978. The impact of tourism on the physical environment. *Annals of Tourism Research* 5(2): 215–237.

COMMON LAND FORUM. 1986. *Common land. The report of the Common Land Forum.* Publication CCP 215. Cheltenham. Countryside Commission. 108pp.

CONVERY, F.J. 1976. *Economics applied to outdoor recreation: an evaluation.* United States Department of Agriculture Forest Service General Technical Report SE9: 108–119.

COLLINS, M. 1986. Foreword to Coalter, F., Long, J. and Duffield, B.S. 1986. *Rationale for public sector investment in leisure.* London. Sports Council and Economic and Social Research Council. 181pp.

COOPERS & LYBRAND ASSOCIATES LTD. 1981. *Sharing does work. Economic and social benefits and costs of direct and joint sports provision.* Sports Council Study 21. London. Sports Council. 35pp.

COUNCIL OF EUROPE. 1976. *European Sport For All charter.* Resolution (76)41 of the Committee of Ministers. Strasbourg. Council of Europe.

COUNTRY LANDOWNERS ASSOCIATION. 1991. *Recreation and access in the countryside: a better way forward.* London. Country Landowners Association. 20pp.

COUNTRYSIDE COMMISSION. 1982. *Countryside issues and action: prospectus of the Countryside Commission 1982.* Publication CCP 151. Cheltenham. Countryside Commission. 15pp.

COUNTRYSIDE COMMISSION. 1985. *National countryside recreation survey: 1984.* Publication CCP 201. Cheltenham. Countryside Commission. 20pp.

COUNTRYSIDE COMMISSION. 1986. *Recreation 2000. A discussion paper on future recreation policies.* Cheltenham. Countryside Commission. 8pp.

COUNTRYSIDE COMMISSION. 1987. *Enjoying the countryside. A consultation paper on future policies.* Publication CCP 225. Cheltenham. Countryside Commission. 32pp.

COUNTRYSIDE COMMISSION. 1987A. *Managing rights of way: an agenda for action.* Publication CCP 273. Cheltenham. Countryside Commission. 24pp.

COUNTRYSIDE COMMISSION. 1988. *Changing the rights of way network. A discussion paper.* Publication CCP 254. Cheltenham. Countryside Commission. 28pp.

COUNTRYSIDE COMMISSION. 1989. *Planning for a greener countryside.* Publication CCP 264. Cheltenham. Countryside Commission.

COUNTRYSIDE COMMISSION. 1989A. *Paths, routes and trails: policies and priorities.* Publication CCP 266. Cheltenham. Countryside Commission. 16pp.

COUNTRYSIDE COMMISSION. 1989B. *Annual report 1988–89.* Cheltenham. Countryside Commission. 52pp.

COUNTRYSIDE COMMISSION. 1991. *An agenda for the countryside.* Publication CCP 336. Cheltenham. Countryside Commission. 11pp.

COUNTRYSIDE COMMISSION. 1991A. *Visitors to the countryside.* Publication CCP 341. Cheltenham. Countryside Commission. 20pp.

COUNTRYSIDE COMMISSION. 1991B. *The Pennine Way management project.* Publication CCP 297. Cheltenham. Countryside Commission. 75pp.

COUNTRYSIDE COMMISSION. 1991C. *Countryside stewardship: an outline.* Publication CCP 346. Cheltenham. Countryside Commission. 5pp.

CULLINGWORTH, J.B. 1976. *Town and country planning in Britain,* 6th. edition. London. George Allen and Unwin Ltd. 287pp.

CULLINGWORTH, J.B. 1988. *Town and country planning in Britain,* 10th. edition. London. Unwin Hyman. 408pp.

CWI, D. 1982. Merit good or market failure: justifying and analysing public support for the arts. pp 59–89 in Mulcahy, K.V. and Swaim, C.A. (editors). *Public policy and the arts.* Boulder, Colo. Westview Press Inc. 332pp.

DARTINGTON AMENITY RESEARCH TRUST. 1974. *Farm recreation and tourism in*

England and Wales. Report to the Countryside Commission, English Tourist Board and Welsh Tourist Board. Publication CCP 83. Cheltenham. Countryside Commission.

DAVIES, E.T. 1971. *Farm tourism in Cornwall and Devon. Some economic and physical considerations.* Report No. 184. Agricultural Economics Unit, University of Exeter. 16pp.

DAVIES, E.T. 1973. *Tourism on Devon farms. A physical and economic appraisal.* Report No. 188. Agricultural Economics Unit, University of Exeter. 39pp.

DAVIES, H.W.E., RAVENSCROFT, N., BISHOP, K. and GOSLING, J.A. 1991. *Operation of the planning system in handling large-scale leisure development proposals.* Final report to the National Economic Development Office. Department of Land Management and Development, University of Reading. 2 volumes: report and conclusions, 70pp. and case studies, 91pp.

DAVIS LANGDON & EVEREST. (editors). 1992. *Spon's architects' and builders' price book,* 117th. edition. London. E. and F.N. Spon. 1043pp.

DAVISON, J. 1988. Urban renaissance. pp 28–29 in Arts Council. 1988. *43rd. annual report and accounts.* London. Arts Council of Great Britain. 120pp.

DAY, M. and DAVIS, M. 1990. Environmental assessment regulations: response of the development sector. *Estates Gazette* 9028: 48–50 (July 14th, 1990).

DENMAN, D.R. 1978. *The place of property. A new recognition of the function and form of property rights in land.* Berkhamstead, Herts. Geographical Publications Ltd. 150pp.

DENMAN, D.R. and PRODANO, S. 1972. *Land use. An introduction to proprietary land use analysis.* London. George Allen and Unwin Ltd. 230pp.

DEPARTMENT OF THE ENVIRONMENT. 1975. *Sport and recreation.* Cmnd. 6200. London. HMSO. 19pp.

DEPARTMENT OF THE ENVIRONMENT. 1977. *Recreation and deprivation in inner urban areas.* London. HMSO. 65pp.

DEPARTMENT OF THE ENVIRONMENT. 1981. *Service provision and pricing in local government: studies in local environmental services.* London. HMSO. 272pp.

DEPARTMENT OF THE ENVIRONMENT, AUDIT INSPECTORATE. 1983. *Development and operation of leisure centres (selected case studies).* London. HMSO. 77pp. plus appendices.

DEPARTMENT OF THE ENVIRONMENT, SCOTTISH EDUCATION DEPARTMENT and WELSH OFFICE. 1987. *Competition in the management of local authority sport and leisure facilities.* Consultation Paper. London. Department of the Environment. 4pp.

DEPARTMENT OF THE ENVIRONMENT and WELSH OFFICE. 1989. *Environmental assessment. A guide to the procedures.* London. HMSO. 64pp.

DOWER, J. 1945. *National parks in England and Wales.* Cmnd. 6628. London. HMSO.

DOWER, M. 1967. *The challenge of leisure.* London. Civic Trust. 72pp.

DOWER, M., RAPOPORT, R., STRELITZ, Z. and KEW, S. 1981. *Leisure provision and people's needs.* London. HMSO. 153pp.

DREWETT, R. 1973. The developers: decision processes. pp 163–193 in Hall, P., Gracey, H., Drewett, R. and Thomas, R. 1973. *The containment of urban England; volume 2: the planning system; objectives, operations, impacts.* London. George Allen and Unwin Ltd. 464pp.

DRUCKER, P.F. 1977. *Management.* London. Pan Books Ltd. 527pp.

DUNN, P. 1991. To lose a valley for the sake of a hosepipe. *The Independent* 9th March 1991: 38.

EAST MIDLANDS COUNCIL FOR SPORT AND RECREATION. 1972. *Recreation in the East Midlands, a regional plan. Part 2, water-based recreation.* Nottingham. East Midlands Council for Sport and Recreation. 72pp.

EDINGTON, J.H. and EDINGTON, M.A. 1986. *Ecology, recreation and tourism.* Cambridge. Cambridge University Press. 200pp.

EDWARDS, J.A. 1989. Historic sites and their local environments. pp 272–293 in Herbert, D.T., Prentice, R.C. and Thomas, C.J. (editors). 1989. *Heritage sites: strategies for marketing and development.* Aldershot, Hants. Avebury. 309pp.

ENEVER, N. 1989. *The valuation of property investments,* 4th. edition. London. The Estates Gazette Ltd. 208pp.

ENGLISH TOURIST BOARD. 1973. *Static holiday caravans and chalets.* London. English Tourist Board. 25pp.

ENGLISH TOURIST BOARD. 1984. *Starting a small guest house or bed and breakfast business.* Development Guide DG 28. London. English Tourist Board. 31pp.

ENGLISH TOURIST BOARD. 1985. *Developing timeshare.* Development Guide DG 34. London. English Tourist Board. 39pp.

ENGLISH TOURIST BOARD. 1988. *Visitors in the countryside.* London. English Tourist Board. 26pp.

ENGLISH TOURIST BOARD. 1990. *Tourism towards the year 2000. A new strategy for England.* Draft for consultation. London. English Tourist Board. 29pp.

ENGLISH TOURIST BOARD. 1991. *Planning for success. A tourism strategy for England 1991–1995.* London. English Tourist Board. 36pp.

ENGLISH TOURIST BOARD and JONES LANG WOOTTON. 1989. *Retail, leisure and tourism.* London. Jones Lang Wootton. 176pp.

EVELY, R.W. 1976. Assessing need. pp 3–10 in Davidson, A.W. and Leonard, J.E. (editors). 1976. *The property development process.* Property Studies in the United Kingdom and Overseas – 7. Reading. Centre for Advanced Land Use Studies, College of Estate Management. 380pp.

FENTEM, P.H. and BASSEY, E.J. 1978. *The case for exercise.* Sports Council Research Working Papers No. 8. London. Sports Council. 8pp.

FENTEM, P.H., BASSEY, E.J. and BLECHER, A. 1979. *Exercise and health: a bibliography of references collected during a literature search for evi-*

dence that exercise is of benefit to health. Nottingham. Department of Physiology and Pharmacology, Medical School, Queen's Medical Centre, Nottingham.

FIELD, B.G. 1988. Public space, private development and community welfare. *Land Development Studies* 5: 139–144.

FIELD, B.G. and MACGREGOR, B.D. 1987. *Forecasting techniques for urban and regional planning.* London. Hutchinson. 238pp.

FLANAGAN, R. and NORMAN, G. 1989. *Life cycle costing.* Oxford. Blackwell Scientific Publishers.

FORESTRY COMMISSION. 1979. *Advice for woodland owners.* Edinburgh. Forestry Commission. 18pp.

FORESTRY COMMISSION. 1981. *Forestry grant scheme.* Edinburgh. Forestry Commission. 8pp.

FORTLAGE, C.A. 1990. *Environmental assessment: a practical guide.* Aldershot, Hants. Gower Publishing Company Ltd. 152pp.

FOTHERGILL, S., MONK, S. and PERRY, M. 1987. *Property and industrial development.* London. Hutchinson. 187pp.

FRASER, W.D. 1984. *Principles of property investment and pricing.* Basingstoke, Hants. Macmillan Education Ltd. 432pp.

FREEMAN, J. 1989. Building without tears. *Leisure Management* 9(3): 98–99.

GILBERT, D. and ARNOLD, L. 1989. Budget hotels. *Leisure Management* 9(2): 61–63.

GLYPTIS, S. 1984. *The home as a leisure centre.* Paper presented at the Leisure Studies Association conference: Leisure: politics, planning and people. Sussex University, Brighton. 4–8 July 1984. 13pp.

GLYPTIS, S. 1989. *Leisure and unemployment.* Milton Keynes. Open University Press. 180pp.

GLYPTIS, S., KAY, T. and DONKIN, D. 1986. *Sport and the unemployed: lessons from schemes in Leicester, Derwentside and Hockley Port.* London. Sports Council. 42pp.

GOODCHILD, R.N. and MUNTON, R.J.C. 1985. *Development and the landowner.* London. George Allen and Unwin Ltd. 210pp.

GORDON, C. 1991. Sustainable leisure. *Ecos* 12(1): 7–13.

GORZ, A. 1987. *Ecology as politics,* 2nd. impression. London. Pluto Press. 215pp.

GOSLING, J.A. 1990. *The Town and Country (Assessment of Environmental Effects) Regulations 1988: the first year of application.* Proposed Environmental Policy Working Paper. Department of Land Management and Development, University of Reading. 46pp plus appendices.

GOULD, B. 1990. *Labour's policies for the countryside.* London. The Labour Party.

GRATTON. C. and TAYLOR, P. 1985. *Sport and recreation: an economic analysis.* London. E. and F.N. Spon Ltd. 261pp.

GRATTON, C. and TAYLOR, P. 1987. *Leisure in Britain.* Hitchin, Herts. Leisure Publications (Letchworth) Ltd. 105pp.

GRATTON, C. and TAYLOR, P. 1987A. Why provide? Management objectives in leisure. *Leisure Management* 7(1): 23–25.

GRATTON, C. and TAYLOR, P. 1988. *Economics of leisure services management.* Harlow, Essex. Longman Group UK Ltd. 167pp.

GRATTON, C. and TAYLOR, P. 1991. Economics of CCT: part 1. *Leisure Management* 11(4): 54–56.

GRATTON, C. and TICE, A. 1987. Leisure participation and health: use of the GHS. *The General Household Survey Newsletter* No. 4: 25–35.

GREATER LONDON AND SOUTH EAST COUNCIL FOR SPORT AND RECREATION. 1971. *A regional strategy for water recreation.* London. Greater London and South East Council for Sport and Recreation. 101pp.

GREENFIELD, M. 1975. NWC: watchdog or lapdog? *Municipal Journal* 83: 21–22.

HAIGH, N. 1987. Collaborative arrangements for environmental protection in Western Europe. pp 366–387 in Enyedi, G., Gijswijt, A.J. and Rhode, B. (editors). 1987. *Environmental policies in East and West.* London. Taylor Graham. 401pp.

HALL, P. 1975. *Urban and regional planning.* Harmondsworth, Middx. Penguin Books Ltd. 312pp.

HALL, P. 1985. The social crisis. *New Society* 74(1195): 320–322.

HALLIDAY, J.E. 1989. Attitudes towards farm diversification. Results from a survey of Devon farms. *Journal of Agricultural Economics* 40(1): 93–100.

HAMPSHIRE COUNTY COUNCIL. 1987. *The future of farming in Hampshire. A response to the discussion paper.* Winchester. Hampshire County Council. 32pp.

HANSON, M. 1989. Partners in a barn stance. *Estates Gazette* 8920: 20–21.

HARRISON, M.L. 1972. Development control. The influence of political, legal and ideological factors. *Town Planning Review* 43: 254–274.

HARTWRIGHT, T.U. 1984. A private enterprise case study – Thorpe Park. *Journal of Planning and Environmental Law* Occasional Papers: Planning for leisure in the countryside 58–69.

HATRY, H.P. 1983. *A review of private approaches for delivery of public services.* Washington, D.C. The Urban Institute Press. 105pp.

HENEAGE COMMITTEE. 1948. *Gathering grounds.* Sub-committee of the Central Water Advisory Committee. London. HMSO.

HENLEY CENTRE. 1991. *Holiday villages: their rationale and economic impact.* A report for Lakewoods Ltd. London. The Henley Centre. 43pp.

HENRY, I.P. 1984. Urban deprivation and the use of social indicators for recreation planning. pp 147–159 in Long, J.A. and Hecock, R.D. (editors). 1984. *Leisure, tourism and social change.* Centre for Leisure Research, Dunfermline College of Physical Education. 223pp.

HENRY, I.P. (editor). 1990. *Management and planning in the leisure industries.* Basingstoke, Hants. Macmillan Education Ltd. 216pp.

HETHERINGTON, J. 1984. Property performance measuring systems. *Estates Gazette* 271(6194): 260–263.

HEWISON, R. 1987. *The heritage industry: Britain in a climate of decline.* London. Methuen London Ltd. 160pp.

HILARY, W. 1984. Leisure and the developer. *Journal of Planning and Environmental Law* Occasional Papers: Planning for leisure in the countryside 90–105.

HILL, E.A. and HEALEY, P. 1985. Local plans for the countryside: the first decade. pp 46–71 in Gilg, A.W. (editor). 1985. *Countryside planning yearbook, volume six.* Norwich. Geo Books. 275pp.

HILL, H. 1980. *Freedom to roam. The struggle for access to Britain's moors and mountains.* Ashbourne, Derbyshire. Moorland Publishing Company Ltd. 139pp.

HILLIAR, P. 1990. The UK leisure market – the Stock Market view. Paper presented at the IBC Conference: *Leisure property into the 1990's.* 24th. January. Inn on the Park. London. 14pp.

HILSON, M.A. 1981. Public health considerations and pollution control aspects. pp 45–57 in Dangerfield, B.J. (editor). 1981. *Recreation: water and land.* Water Practice Manuals 2. London. The Institution of Water Engineers and Scientists. 336pp.

HOAR, D.W. and SWAIN, H.T. 1977. Control of total cost. pp 25–40 in Croome, D.J. and Sherratt, A.F.C. (editors). 1977. *Quality and total cost in buildings and services design.* Lancaster. The Construction Press Ltd. 168pp.

HOBHOUSE, A. 1947. *Report of the National Park Committee (England and Wales).* Cmnd. 6628. London. HMSO.

HOOKWAY, R. 1984. A challenging horizon. *Journal of Planning and Environmental Law* Occasional Papers: Planning for leisure in the countryside: 1–19.

HOUSE OF LORDS SELECT COMMITTEE. 1973. *Sport and leisure.* Parliamentary Papers, House of Lords IX (Miscellaneous 2). 1st. report 16pp; 2nd. report 609pp.

ILBERY, B.W. 1987. The development of farm diversification in the UK: evidence from Birmingham's urban fringe. *Journal of the Royal Agricultural Society of England* 148: 21–35.

JANUARIUS, M. 1991. Max Pearce. The managing director of Lakewoods talks to Mary Januarius about the Cotswold Water Park and rural development issues. *Leisure Management* 11(10): 24–27.

JENKINS, A. 1974. *The case for squash: its growth, development and prospects.* Leigh-on-Sea, Essex. A. Jenkins in association with the Squash Rackets Association. 68pp.

JENKINS, C. and SHERMAN, B. 1979. *The collapse of work.* London. Eyre Methuen. 182pp.

JOHN, G. and HEARD, H. (editors). 1981. *Handbook of sports and recreational building design, volume 1: ice rinks and swimming pools.* London. The Architectural Press Ltd. 155pp.

JOHN, G. and HEARD, H. (editors). 1981A. *Handbook of sports and recreational building design, volume 2: indoor sports.* London. The Architectura Press Ltd. 214pp.

JOHN, G. and HEARD, H. (editors). 1981B. *Handbook of sports and recreational building design, volume 3: outdoor sports.* London. The Architectural Press Ltd. 176pp.

JOHN, G. and HEARD, H. (editors). 1981C. *Handbook of sports and recreational building design, volume 4: sports data.* London. The Architectural Press Ltd. 134pp.

JOHNSON, S. 1990. Jilted by the City. *Leisure Property, a CSW Supplement,* 31st. May 1990: 9.

JOINT CENTRE FOR LAND DEVELOPMENT STUDIES. 1985. *Ploughing footpaths and bridleways.* Publication CCP 190. Cheltenham. Countryside Commission. 104pp.

JONES, D. 1989. *Financial vulnerability in farming. A study of North West farms.* Bulletin 220. Department of Agricultural Economics, University of Manchester. 32pp.

KEITH, N. 1989. Getting out of a hole. *Leisure; a CSW Supplement.* 7th. December 1989: 16–17.

KEMPNER, T. (editor). 1980. *A handbook of management,* 3rd. edition. Harmondsworth, Middx. Penguin Books Ltd. 446pp.

KOTLER, P. 1984. *Marketing management: analysis, planning and control,* 5th. edition. Englewood Cliffs, New Jersey. Prentice-Hall. 794pp.

KOUSKOULAS, V. and KOEHN, E. 1974. Predesign cost-estimation function for buildings. *Proceedings of the American Society of Civil Engineers; Journal of the Construction Division* 100(CO4): 589–604.

LABOUR PARTY. 1990. *An earthly chance. Labour's programme for a cleaner greener Britain, a safer sustainable planet.* London. Labour Party. 37pp.

LE GRAND, J. 1982. *The strategy of equality. Redistribution and the social services.* London. George Allen and Unwin Ltd. 192pp.

LE GRAND, J. and ROBINSON, R. 1984. Privatisation and the welfare state: an introduction. pp 1–4 in Le Grand, J. and Robinson, R. (editors). 1984. *Privatisation and the welfare state.* London. George Allen and Unwin Ltd. 233pp.

LIBERAL DEMOCRATS. 1990. *A thriving countryside. Liberal Democrat policies for rural affairs and agriculture.* Federal Green Paper No. 18. Dorchester, Dorset. Liberal Democrat Publications. 40pp.

LICHFIELD, N. 1956. *Economics of planned development.* London. Estates Gazette Ltd. 460pp.

LICHFIELD, N. and DARIN-DRABKIN, H. 1980. *Land policy in planning.* London. George Allen and Unwin Ltd. 321pp.

LIKIERMAN, A. 1979. The financial and economic framework for nationalised industries. *Lloyds Bank Review* 134: 16–32.

LONG, J.A. and HECOCK, R.D. 1984. Editorial foreword. pp vii-x in Long, J.A. and Hecock, R.D. (editors). 1984. *Leisure, tourism and social change.*

238   *References*

Centre for Leisure Research, Dunfermline College of Physical Education. 223pp.

LOWE, P., COX, G., MacEWEN, M., O'RIORDAN, T. and WINTER, M. 1986. *Countryside conflicts. The politics of farming, forestry and conservation.* Aldershot, Hants. Gower Publishing Company Ltd. 378pp.

LUMBY, S.P. 1984. *Investment appraisal.* Wokingham, Berks. Van Nostrand Reinhold (UK) Co Ltd. 323pp.

LYNCH, J. 1989. Leisure construction: a new direction. *Leisure Management* 9(1): 85–86.

McCORMACK, T. 1971. Politics and leisure. *International Journal of Comparative Sociology* 12(3): 168–181.

MacEWEN, A. and MacEWEN, M. 1982. *National parks: conservation or cosmetics?* London. George Allen & Unwin (Publishers) Ltd. 314pp.

McINTOSH, P. 1966. *Sport in society.* London. Watts.

McINTOSH, P. and CHARLTON, V. 1985. *The impact of Sport For All policy 1966–1984 and a way forward.* London. Sports Council. 215pp.

MACKAY, K.J. and CROMPTON, J.L. 1988. A conceptual model of consumer evaluation of recreation service quality. *Leisure Studies* 7(1): 41–49.

McMILLAN SCOTT, E. 1990. Growth of tourism: can the environment cope? pp 240–246 in Quest, M. (editor). 1990. *Horwath book of tourism.* London. The Macmillan Press Ltd. 264pp.

McNAMARA, P.F. 1983. Towards a classification of land developers. *Urban Law and Policy* 6: 87–94.

MASLOW, A.H. 1970. *Motivation and personality.* 2nd edition. New York. Harper and Row Publishers, Inc. 369pp.

MAYNARD, A. 1983. Privatizing the National Health Service. *Lloyds Bank Review* 148: 28–41.

MEGGITT, D.I. 1980. Modelling life-cycle costs for new cities overseas. *Asset Management News* 7: 1–4.

MENDOZA, L. n.d. *Some leisure definitions and development parameters.* Hove, Sussex. The Lionel Mendoza Partnership. 14pp.

MERCER, D. 1973. The concept of recreational need. *Journal of Leisure Research* 5: 37–50.

MICHEL, C. 1979. *The cost of hospitalisation. Micro-economic approach to the problems involved.* Social Policy Series No. 39. Brussels. Commission of the European Communities. 64pp.

MIDDLETON, V.T.C. 1990. *New visions for independent museums in the UK.* Chichester, W. Sussex. Association of Independent Museums. 80pp.

MILES, C.W.N. 1986. *Running an open house: a guide to the public opening of houses.* London. Surveyors Publications. 36pp.

MILES, C.W.N. and SEABROOKE, W. 1977. *Recreational land management.* London. E. and F.N. Spon Ltd. 147pp.

MILLER, G. 1991. Environmental economics – the bridge between community need and market forces? Paper presented at the Leisure Studies Association annual conference: *Leisure in the 1990's: rolling back the welfare state*. University of Ulster at Jordanstown, 12–15 September 1991.

MILLS, A.S. 1985. Participation motivations for outdoor recreation: a test of Maslow's theory. *Journal of Leisure Research* 17(3): 184–199.

MINISTRY OF AGRICULTURE, FISHERIES AND FOOD. 1989. *Agriculture in the UK 1988*. London. HMSO.

MINISTRY OF HOUSING AND LOCAL GOVERNMENT. 1970. *Development Plan Manual*. London. HMSO.

MISHAN, E.J. 1976. *Elements of cost-benefit analysis*, 2nd. edition. London. George Allen and Unwin Ltd. 151pp.

MONTAGU EVANS. 1990. *Preliminary assessment. Summary of the differences between the alternative development schemes*. London. Environmental Assessment Group, Montagu Evans.

MONTAGU EVANS. 1990A. *Proposed leisure centre, Stoke Park; summary environmental report*. London. Environmental Assessment Group, Montagu Evans. 15pp.

MONTAGU OF BEAULIEU, LORD. 1968. *The gilt and the gingerbread*. London. Sphere Books Ltd. 192pp.

MOORE, V. 1987. *A practical approach to planning law*. London. Financial Training Publications Ltd. 347pp.

MOYNIHAN, C. 1987. *The future direction of national sports policies*. Open letter to John Smith, Chair of the Sports Council. London. Department of the Environment. 3pp.

MULCAHY, K.V. 1982. The rationale for public culture. pp 33–58 in Mulcahy, K.V. and Swaim, C.R. (editors). 1982. *Public policy and the arts*. Boulder, Colo. Westview Press Inc. 332pp.

MURPHY, P.E. 1985. *Tourism: a community approach*. London. Methuen and Co. Ltd. 200pp.

NATIONAL FARMERS UNION. 1973. *Farm tourism in Wales*. A report of the NFU Council for Wales. London. National Farmers Union.

NEULINGER, J. 1982. Leisure lack and the quality of life: the broadening scope of the leisure professional. *Leisure Studies* 1(1): 53–63.

NEWMAN, M. and ROBERTS, E. 1989. Taking the Mickey out of the USA. *Leisure; a CSW Supplement*, 7th. December. 1989: 37–39.

NG, Y. 1983. *Welfare economics*. Basingstoke, Hants. Macmillan Publishers Ltd. 333pp.

NORTHERN COUNCIL FOR SPORT AND RECREATION. 1982. *A strategy for sport and recreation in the Northern region*. Newcastle-upon-Tyne. Northern Council for Sport and Recreation. 11pp.

OLIVER, D.M. 1984. *Managing inland water for leisure and recreation*. Paper presented at an Institute of Leisure and Amenity Management, Southern Region, seminar. Hampshire College of Agriculture. 15th. November 1984. 15pp.

OLIVER, D.M. 1985. Managing inland water for leisure and recreation – an example from southern England. *The Environmentalist* 5(3): 171–178.

ONTARIO MINISTRY OF TOURISM AND RECREATION. 1983. *Cost revenue squeeze: managing recreation services to increase revenues and decrease expenditures.* Toronto, Canada. Government of Ontario.

ORGANISATION FOR ECONOMIC CO-OPERATION AND DEVELOPMENT. 1980. *The impact of tourism on the environment.* General report. Paris. OECD. 148pp.

ORGANISATION FOR ECONOMIC CO-OPERATION AND DEVELOPMENT, GROUP ON URBAN AFFAIRS. 1986. *Managing and financing urban services.* Draft final report. Paris. OECD. 111pp.

PAFFENBERGER, R.S., HYDE, R.T., WING, A.L. and CHUNG-CHENG, H. 1986. Physical activity, all-cause mortality, and longevity of college alumni. *The New England Journal of Medicine* 314(10): 605–613.

PAICE, C. 1988. Efficiency beats all the exotic alternative ideas. *Farmers Weekly* 109(2): 30–31.

PARKER, D.J. and PENNING-ROWSELL, E.C. 1980. *Water planning in Britain.* London. George Allen and Unwin Ltd. 277pp.

PARKER, R.R. 1987. *Land: new ways to profit.* London. Country Landowner's Association. 156pp.

PARKER, S. 1976. *The sociology of leisure.* London. George Allen and Unwin Ltd. 157pp.

PARKINSON, M. 1989. The Thatcher Government's urban policy, 1979–1989. *Town Planning Review* 60(4): 421–440.

PATMORE, J.A. 1983. *Recreation and resources.* Oxford. Basil Blackwell Publisher Ltd. 280pp.

PEACOCK, A. and GODFREY, C. 1975. Public provision of museums and galleries: the economic issues. pp 128–140 in Searle, G.A.C. (editor). 1975. *Recreational economics and analysis.* Harlow, Essex. Longman Group Ltd. 190pp.

PEARCE, D.G. 1985. Tourism and environmental research: a review. *International Journal of Environmental Studies* 25: 247–255.

PEARCE, D.G. 1989. *Tourist development,* 2nd edition. Harlow, Essex. Longman Group UK Ltd. 341pp.

PEARCE, D.W. and TURNER, R.K. 1990. *Economics of natural resources and the environment.* Hemel Hempstead, Herts. Harvester Wheatsheaf. 378pp.

PIGRAM, J.J. 1983. *Outdoor recreation and resource management.* Beckenham, Kent. Croom Helm Ltd. 262pp.

POTIRIADIS, M. 1990. The importance of good design: leisure design: suitable for purpose? Paper presented at the IBC Conference: *Leisure property into the 1990's.* 24th. January 1990. Inn on the Park, London. 6pp.

PUNTER, J.V. 1986. Aesthetic control within the development process: a case study. *Land Development Studies* 3: 197–212.

PURKIS, H.J., HOW, R.F.C., HOOPER, N.J. and POOLE, M.T. 1977. *Occupancy costs of offices.* Current Paper 44/77. Garston, Watford. Building Research Establishment. 7pp.

PYE-SMITH, C. and HALL, C. (editors). 1987. *The countryside we want: a manifesto for the year 2000.* Bideford, Devon. Green Books. 134pp.

RATCLIFFE, J. n.d. *Leisure and related development.* Occasional Paper PD13. Department of Estate Management, Polytechnic of the South Bank.

RAVENSCROFT, N. 1988. *An examination of the methods of resource allocation applicable to the public provision of leisure facilities in Great Britain.* Unpublished PhD thesis, University of Reading. 538pp.

RAVENSCROFT, N. and STABLER, M.J. 1986. *The finance of sport in a rural district; voluntary sector case studies: report of survey.* Department of Economics, University of Reading.

READE, E. 1987. *British town and country planning.* Milton Keynes. Open University Press. 270pp.

REES, J.A. 1990. *Natural resources: allocation, economics and policy,* 2nd edition. London. Routledge. 499pp.

RIDDICK, C.C. and DANIEL, S.N. 1984. The relative contribution of leisure activities and other factors to the mental health of older women. *Journal of Leisure Research* 16(2): 136–148.

ROBERTS, J. 1979. *A review of the commercial sector in leisure.* London. Sports Council and Social Science Research Council. 66pp.

ROBERTS, J. 1990. Development finance for leisure projects. Paper presented at the IBC Conference: *Leisure property into the 1990's.* 24th. January 1990. Inn on the Park, London.

ROBERTS, K. 1978. *Contemporary society and the growth of leisure.* London. Longman Group Ltd. 191pp.

RMC GROUP plc. 1987. *A practical guide to restoration,* 2nd. edition. Feltham, Middx. RMC Group plc. 83pp.

RUEGG, R.T., McCONNAUGHEY, J.S., SAV, G.T. and HOCKENBERY, K.A. 1978. *Life-cycle costing. A guide for selecting energy conservation projects for public buildings.* National Bureau of Standards Building Science Series 113. Washington, D.C. United States Department of Commerce. 70pp.

RUSSELL, R.V. 1987. The relative contribution of recreation satisfaction and activity participation to the life satisfaction of retirees. *Journal of Leisure Research* 19(4): 273–283.

SCARMAN, L.G., Baron. 1981. *The Brixton disorders 10–12 April 1981; report of an inquiry.* Cmnd. 8427. London. HMSO. 168pp.

SCOTT, MJ. 1942. *Report of the committee on land utilisation in rural areas.* Cmnd. 6378. London. HMSO.

SCOTTISH SPORTS COUNCIL. 1979. *A study of sports centres and swimming pools. Main report: a question of balance.* 2 volumes. Edinburgh. Scottish Sports Council. 290pp. plus appendices.

SEABROOKE, W. 1989. *Land management or great expectations?* Transcript of an inaugural lecture given at Portsmouth Polytechnic, 6th. December 1989. 20pp.

SEBAG-MONTEFIORE, R.O. 1991. *Land values.* Paper presented at the joint

University of Reading/College of Estate Management Land Agents Conference and Seminar. University of Reading, January 16th. 1991.

SECRETARIES OF STATE FOR THE ENVIRONMENT, TRADE AND INDUSTRY, HEALTH, EDUCATION AND SCIENCE, SCOTLAND, TRANSPORT, ENERGY AND NORTHERN IRELAND, THE MINISTER OF AGRICULTURE, FISHERIES AND FOOD AND THE SECRETARIES OF STATE FOR EMPLOYMENT AND WALES. 1990. *This common inheritance. Britain's environmental strategy.* Cmnd. 1200. London. HMSO. 291pp.

SECRETARY OF STATE FOR THE ENVIRONMENT. 1977. *Policy for the inner cities.* Cmnd. 6845. London. HMSO.

SEELEY, I.H. 1973. *Outdoor recreation and the urban environment.* London. The Macmillan Press Ltd. 235pp.

SEELEY, I.H. 1992. *Public works engineering.* Basingstoke. Macmillan Education Ltd.

SHOARD, M. 1987. *This land is our land. The struggle for Britain's countryside.* London. Paladin Grafton Books. 592pp.

SIBLEY, N. 1990. Setting the parameters for successful leisure developments. Paper presented at the IBC Conference: *Leisure property into the 1990's.* 24th. January 1990. Inn on the Park, London.

SIDAWAY, R. 1988. *Sport, recreation and nature conservation.* Study 32. London. Sports Council. 98pp.

SIMMIE, J. 1981. *Power, property and corporatism.* London. The Macmillan Press Ltd. 351pp.

SLEE, W. 1987. *Alternative farm enterprises.* Ipswich. Farming Press. 210pp.

SMITH, D.L. 1974. *Amenity and urban planning.* London. Crosby Lockwood Staples. 198pp.

SMITH, R. 1988. Development economics: leisure buildings 3: initial cost estimating. *The Architect's Journal* 188(50): 55–64.

SOLESBURY, W. 1975. Ideas about structure plans: past, present and future. *Town Planning Review* 46 (July 1975).

SOUTH, R. 1978. *Crown, college and railways.* Buckingham. Barracuda Books Ltd. 136pp.

SOUTH WESTERN COUNCIL FOR SPORT AND RECREATION. 1971. *Water recreation strategy.* Crewkerne. South Western Council for Sport and Recreation. 34pp.

SPORTS COUNCIL. 1972. *Provision for sport.* London. HMSO. 12pp.

SPORTS COUNCIL. 1974. *Annual report 1973–74.* London. Sports Council. 57pp.

SPORTS COUNCIL. 1975. *Sports centre running costs: a study of running costs of six multi-purpose sports centres.* London. Sports Council. 43pp.

SPORTS COUNCIL. 1976. *Recreation facilities: some ideas on reducing their deficits.* London. Sports Council. 14pp.

SPORTS COUNCIL. 1977. *Recreational use of church buildings.* Technical Unit for Sport Design Note 7. London. Sports Council.

SPORTS COUNCIL. 1980. *A low cost sports hall: an account of the building and subsequent use of Rochford Sports Centre.* Technical Unit for Sport Design Note 8. London. Sports Council. 15pp.

SPORTS COUNCIL. 1982. *Sport in the community: the next ten years.* London. Sports Council. 53pp.

SPORTS COUNCIL. 1982A. *Sports Council announce £5m standardised sports hall scheme.* Press Release SC164. London. Sports Council. 2pp.

SPORTS COUNCIL. 1985. *Annual report 1984/85.* London. Sports Council. 57pp.

SPORTS COUNCIL. 1985A. *SASH design guide 1.* London. Sports Council. 184pp.

SPORTS COUNCIL. 1990. *A countryside for sport. Towards a policy for sport and recreation in the countryside: a consultation.* London. Sports Council. 15pp.

STABLER, M.J. 1982. *The finance of sport and recreation in a rural district: voluntary sector study.* Summary report to the Sports Council, 1st. draft. Department of Economics, University of Reading. 12pp.

STABLER, M.J. 1990. Financial management and leisure provision. pp 97–126 in Henry, I.P. (editor). 1990. *Management and planning in the leisure industries.* Basingstoke, Hants. Macmillan Education Ltd. 216pp.

STEPHENS, B. 1983. The economics of design – a standardised approach to sports halls. pp 121–126 in Sports Council. 1983. *Finance and sport – accounting for the future.* Recreation Management Seminar Report. London. Sports Council. 225pp.

STEPHENSON, T. 1989. *Forbidden land. The struggle for access to mountain and moorland.* Manchester. Manchester University Press. 243pp.

STEVENS, T. 1990. Greener than green. *Leisure Management* 10(9): 64–66.

STONE, P.A. 1980. *Building design evaluation: costs-in-use.* London. E. and F.N. Spon Ltd. 235pp.

STONOR, W. 1983. Low cost sports building. *Leisure Management* 3(3): 23, 32–34.

STROUD, H.B. 1983. Environmental problems associated with large recreational subdivisions. *Professional Geographer* 35(3): 303–313.

TERRY, B. 1990. Financing leisure development and property. Paper presented at the IBC Conference: *Leisure property into the 1990's.* 24th. January 1990. Inn on the Park, London.

THOMAS, K. 1991. CCT: round one. *Leisure Management* 11(3): 46–47.

THOMAS, K. 1991A. CCT: worth the effort? *Leisure Management* 11(9): 41–42.

THORNLEY, A. 1991. *Urban planning under Thatcherism: the challenge of the market.* London. Routledge. 253pp.

TINSLEY, H.E.A. and JOHNSON, T.L. 1984. A preliminary taxonomy of leisure activities. *Journal of Leisure Research* 16(3): 234–244.

TOMLINSON, A. 1979. *Leisure and the role of clubs and voluntary groups.* London. Sports Council and Social Science Research Council. 60pp.

TORKILDSEN, G. 1986. *Leisure and recreation management,* 2nd. edition. London. E. and F.N. Spon Ltd. 525pp.

TORKILDSEN, G. 1986. *Harlow Sportcentre: the changing and changeless sports centre.* Harlow. George Torkildsen, Leisure and Recreation Management Consultant. 143pp. plus appendices.

TRAVIS, A.S. 1979. *The state and leisure provision.* London. Sports Council and Social Science Research Council. 39pp.

TURNER, A. 1991. *Building procurement.* Basingstoke. Macmillan Education Ltd. 179pp.

UTHWATT REPORT. 1942. *Report of the expert committee on compensation and betterment.* Cmnd. 6386. London. HMSO.

VASALLO, I. and DELALANDE, S. 1980. An overview of the present state and the future prospects. pp 21–47 in Organisation for Economic Co-operation and Development. 1980. *The impact of tourism on the environment.* General report. Paris. OECD. 148pp.

VEAL, A.J. 1979. *Six examples of low cost sports facilities.* Sports Council Study 20. London. Sports Council. 50pp.

VEAL, A.J. 1982. *Planning for leisure: alternative approaches.* Papers in Leisure Studies No. 5. Polytechnic of North London. 55pp.

WAGAR, J.A. 1964. *The carrying capacity of wild lands for recreation.* Monograph No. 17. Logan. USDA Forest Service. 24pp.

WALKER, A. 1984. The political economy of privatisation. pp 19–44 in Le Grand, J. and Robinson, R. (editors). 1984. *Privatisation and the welfare state.* London. George Allen and Unwin Ltd. 233pp.

WALSH, R.G. 1986. *Recreation economic decisions: comparing benefits and costs.* State College, Pennsylvania. Venture Publishing Inc. 637pp.

WELSH WATER AUTHORITY. 1980. *A strategic plan for water-space recreation and amenity.* Cardiff. Welsh Water Authority. 194pp.

WOLFENDEN COMMITTEE ON SPORT. 1960. *Sport and the community.* London. Central Council of Physical Recreation. 135pp.

INDEX

Ability to pay 49, 50
(see also *willingness to pay*)
Access 141
(see also *trespass: mass*)
   agreements 138, 141, 142, 165,
      223, 225
   charges 138, 157
   charter 140–1
   to common land 28
   to the countryside 10, 12, 13,
      28, 64, 123, 128, 133, 144,
      146
   demand 139
   Government policy 127
   to leisure facilities 181, 209, 210
   movement 139, 140
   in National Parks 125–6
   to open country 10, 123, 126,
      146, 162
   pressure 142
   to reservoirs and water courses
      152–3, 157
   rights 10, 26, 28, 64, 138–145,
      157
Access to Commons and Open
   Country Bill 1983 145
Access to Mountains Act 1939 12,
   140
Access to Mountains (Scotland) Bill
   1884 139
Accommodation 18, 148, 150, 193
(see also *farm diversification)*
Accounting
(see also *investment; investor's
   yield)*

discounted cash flow 96
internal rate of return 96
net present value 96
payback method 95
profit/capital ratio 112
rate of return 54, 95
ratios 83, 112
Action Areas 118
Addison Committee 140
Agriculture Act 1947 123
Agriculture Act 1986 147
Aims and objectives 76–9, 90, 179
Allotments 167, 214
Alton Towers 74, 218
Amphitheatres 205, 209
ARC Properties Ltd 200, 208, 209
Areas of Great Landscape Value
   124
Areas of Outstanding Natural
   Beauty 125, 185
Article 4 Directions 205, 207
Art galleries 58, 78
Arts 8, 31, 47, 54, 60
(see also *public arts policies)*
   centres 38
Audit Commission 94, 178
Audit Inspectorate 83

Balance of Payments 97
Barlow Report 123
Basingstoke Canal 158, 191
Berkshire Recreation Subject Plan
   131
Birdwatching 188
Board of Trade 14

Bovis Construction Ltd 87–8
Bowling 57, 73, 211, 215
Boxing 7
Boy Scouts 11
Brecon Beacons National Park 125
British Olympic Committee 9
(see also *Olympic games*)
British Rail Property Board 215–7
British Waterways 97–8, 154–8
British Workers' Sports Federation
    140
Bryce, J. 139
Building regulations 119
Bureau of Outdoor Recreation 13
Business Expansion Scheme 104

Camping grounds 143, 150
Canoeing 152, 153, 154
Capitalism
    markets 34, 47
    system 26
Car parking 54, 143, 158, 161, 212
Carrying capacity 194
Cash flow 91
Catholic Church 8
Center Parcs 208, 222
Central Council for Physical
    Recreation 9, 11 13
Chessington World of Adventures
    218
Church of England 162
Cinemas 38, 39 40, 49, 57, 83
Civic Trust 142
Civil unrest 16
Clawson, M. 5, 74, 98
Clubs (see voluntary clubs)
Collectivism
    state provision 34
    system 26, 34
Commercial
    leisure development proposals
        62, 199
    provision 18, 56
    sector 19, 20, 38, 54–8, 76, 95,
        112, 171

(see also *ideology of the market;
    private sector*)
sporting and leisure activities 59
Committee for the Conservation of
    Nature and Natural
    Environments 182
Common land 138–43
(see also *access to common land;
    urban commons*)
    registration 142
Common Land Forum 145–6
Commons, Footpaths and Open
    Spaces Preservation Society
    139
Commons Registration Act
    1965 142
Community participation 17, 177,
    195
(see also *participation*)
Community Land Act 1974 130–1
Compulsory competitive tendering
    20, 21, 38, 54, 58, 94, 105,
    113–14, 165, 168, 178–9, 215
Congestion 13, 192, 193, 204, 217
(see also *traffic congestion*)
Conservative Party 16, 20, 120,
    128, 134, 157, 162, 172
Construction 43, 105
    building contracts 105
    Design and Build contract
        105–6, 215
    JCT contract 105–6
Consumer surplus 93, 97–8
Consumption 71
    barriers to 65
    of leisure facilities 69
    services 49
cost 79, 81
(see also *costs in use; life cycle
    cost; social cost*)
    annual cost model 90
    appraisal 79, 83
    capital 82–8, 95
    of development and operation 95
    efficiency 83, 85

environmental 82, 182, 198
explicit 81
of facilities 167
implicit 81, 88
of land and construction 76
of leisure facilities 84
maintenance 84
operating 76, 82–5
opportunity 81, 97, 112
recurrent 82–8
saving 83
staff 84, 88
of transfer 36
Cost benefit analysis 97–9, 102,
    170, 182, 226
Costs in use 82–4
Cotswold District Council 208–10
Cotswold Water Park 207–9
Council for the Protection of Rural
    England 161, 214
Council of Europe 182
Country
    parks 11, 13, 19, 64, 143–4,
        208, 225
    sports 7, 163, 190
Country Landowners' Association
    146
Countryside Act 1968 13, 128,
    142–4
Countryside Commission 13, 14,
    31, 53, 84, 103, 127–8,
    143–7, 160–2, 171, 192
(see also *Countryside Stewardship
    Scheme*)
Countryside
(see also *access to the countryside;
    development in the
    countryside*)
    conservation 11
    future 117, 161–3
    recreation facilities 128
    (see also *demand*)
Countryside Policy Review Panel
    147

Countryside Stewardship Scheme
    146
Cycling 50

Demand 2, 35, 42, 49, 65, 71, 92,
    93, 152, 207
(see also *access: demand*)
    analysis 94
    consumer 49
    for countryside recreation 138–9
    derived 49
    economist's concept 72
    effective 73
    elasticity 92–3
    forecasting 49, 74, 93
    influences on 73
    for land and property 36, 49, 50,
        125
    for land suitable for golf course
        development 19
    latent 69, 75, 104
    for leisure facilities 165, 199, 200
    option 168–71
    patterns of 73
    for recreation facilities 11, 50,
        128, 148
    for travel 13
Democratisation of urban sport and
    recreation 10
Demonstration farms 205
Department of Education and
    Science 178
Department of Employment 16, 22,
    31
Department of the Environment 14,
    15, 17, 93, 129–36, 156, 160,
    178, 185, 195, 222
    Circulars 130–2, 147
Department of Trade and Industry
    16, 22, 31
Developer's
    profit 107–14
    yield 108, 111
Development 41

(see also *commercial leisure development proposals; cost of development and operation; economic development; evaluation; facility development; financing leisure development; intention to develop; large scale leisure and tourism developments; leisure development; market for development land; planning for recreation development; property developers; property development; public leisure development; recreation development; social impact of development; tourism development; tourism: sustainable development*)
charge 125
control 116–8, 124, 129, 134, 138, 183–4, 224
in the countryside 116, 123, 161, 219, 222
legal definition 41, 118
intention 50, 56, 58, 64
of leisure facilities 14, 68, 201
of public open space 201
of squash 19
of tourist facilities 14, 30, 60, 132, 161, 183, 191–14
motives 65
plans 116–18, 124, 127–35, 224–5
pressure 117
process 41–50, 91, 105, 107, 114, 116, 196–9, 208, 226
proposals 199, 222
speculative 35, 103
value 122–5, 130
visual impact 191
Development Land Tax Act 1976 130
Development Plan Manual 129

Development of Tourism Act 1969 14
Disabled people 15
Disney 60–1
Don Valley Linear Park 61
Dower, J. 12, 141

Ecological impact of recreation and tourism 188
Economic
benefits of tourism 198
development 60
efficiency 84
growth 65
impact of recreation and tourism 209, 226
implications of facility provision 85
regeneration 60, 132
Economics of sport and recreation 168
(see also *ability to pay; consumer surplus; externality; free rider concept; market failure; merit goods; producers' goods; public goods; welfare economics*)
Education Act 1870 9
Education Act 1944 12, 51
EEC Directive 85/337 184–5
Effectiveness 113
of service delivery 95
Efficiency 53, 113, 181
audits 21
financial 95
indicators 113
Electra Leisure 40, 104
Elitism 10
Employment
generation 65
opportunities 204, 209
Enforcement notice 119
English Heritage 31, 54, 93
English Tourist Board 14, 16, 22, 31, 84, 85, 162, 221

Environmental
(see also *international
    environmental policies;
    tourism:environmental effects*)
  assessment 182–4, 187, 196–7,
      213
  assessment procedures 184–5
  awareness 183, 195
  benefits 182, 198
  costs 182
  damage 190, 197, 210
  degradation 192, 197
  effects 184–7
  impact 153, 182, 184–96,
      209, 210
  impact assessment 184
  impact of coastal development
      192
  impact of recreation and tourism
      development 187, 197
  impact of recreation and tourism
      on historic sites 187
  impact of recreation, sport and
      tourist activities 187–9, 192
  improvement 187, 209
  preservation 121
  statements 184–6, 195–7,
      209–13, 220–6
  stress 191–3
  sustainability 194–5
Environmentally sensitive 193
Equity 35, 167
European Community 183
European Council of Ministers
    183–4
European Leisure and Recreation
    Association 1
European Sport For All Charter 4
Evaluation 42, 94
  of a development project 108
  of the development process 107
  of the operation of a leisure
      facility 112
Externality 53, 116–17, 123,
    168–75

Facility 4
  complimentary 92
  development 13, 63
  indoor 38, 54
  provision 177
  (see also *provision of facilities*)
  public sector 179
Farm
  diversification 19, 57, 138,
      147–51
  (see also *redundant barns*)
  non-agricultural development
      149, 159
  tourism 59, 148
Farm Diversification Grant Scheme
    30, 147, 159, 161
Feasibility 64–5
(see also *accounting*)
  criteria 67
  of a proposed development 75,
      91
  study 42
Financial
(see also *accounting; performance*)
  analysis 81, 95
  appraisal 76, 79
  evaluation 112
  implications of leisure provision
      84–5
  markets 37
  objectives 76
  performance 84
  plan 79
  viability 76, 86, 209
Financial Times Actuaries Leisure
    Index 38, 58
Financing leisure development 61
Fishing 7, 50, 152–6, 165, 190,
    199, 202
Fitness clubs and centres (see
    health and fitness)
Forecasting 36, 75
(see also *performance*)
  techniques 67, 74

Forestry Commission 28, 31, 62, 126, 142, 153
    grants 62
Free rider concept 170
Free time 7
Freehold interest (see property03:interests in)
Funding 76, 79, 103–4, 178
(see also *Electra Leisure*)

Garden cities 121
(see also *Howard, E.*)
General Development Orders 119–20
Gifted athletes 15
Girl Guides 11
Gloucestershire County Council 208, 211
Goals Achievements Matrix 99
Golf clubs and courses 38–40, 47, 51, 63, 73, 108, 150–1, 161, 165, 225
Goodwood Estate 64
Government
(see also *collectivism; state*)
    grants and allowances 62
    (see also *grant aid*)
    policy 17, 45, 151, 207
    (see also *access :government policy*)
    role in society 11
    support for tourism 132
Granada Group plc 200, 208–11
(see also *Lakewoods Ltd; Park Hall Leisure; Somerford Keynes Holiday Village*)
Grant aid 13, 14, 85, 103, 143, 177
Gravity models 74
Greater London Council 70
Green Belts 124, 203, 206, 220, 225
Guildford
    Borough Local Plan 201, 214, 220
    leisure centre 164, 211

Health and fitness 9, 169, 175
(see also *physical fitness*)
    boom 20
    centre 215, 225
    suite 211
Heneage Committee 153
Hill, O. 121
Historic houses 19, 58, 62, 138
(see also *listed buildings; visitors :impact on historic sites*)
Hobhouse, Sir A. 12, 141
Holiday
    complexes 185
    resorts 206
    villages 161, 195, 208–10, 222
Horse racing 7, 8
Hotels 24, 38, 39, 59, 104, 108, 158, 195
House of Lords Select Committee on Sport and Leisure 14, 156
Housing Act 1924 122
Housing and Town Planning Act 1909 121
Howard, E. 121
(see also *garden cities*)
Howell, D. 15, 205
Hunting (see country sports)

Ice rinks 84, 205, 211, 214, 220
Ideology
    of the market 46
    of welfare 46
Income 90
Independent museum sector (see museums)
Informal
    access (see access)
    recreation (see recreation)
Inner city areas 15
Input-output relationships 100, 102
Input measures 79, 180
Institution of Water Engineers and Scientists 153
Intention to develop leisure facilities 58

International
  competition 9
  environmental policies 183
Investment
  appraisal 95
  method of valuation 108
Investor's yield 111

Jogging (see running)
Joint provision of facilities 83
(see also *schools*)

Kennet and Avon Canal 155, 158–9
Keynes, J.M. 123
Kinder Scout (see trespass: mass)

L & R Leisure 218
(see also *Royalty and Empire*)
Labour Party 123, 128, 130, 141,
  162, 182
Lake District National Park 1, 152,
  153, 157
Lakewoods Ltd 200, 211, 222
(see also *Somerford Keynes
  Holiday Village*)
Land 23, 24, 26
(see also *supply of land*)
  indivisibility 37
  legal definition 23
  markets 199
  owned by water companies 157
  policy 29
  source of wealth and power 26
  values 36, 121
Land Commission Act 1967 128
Landed property 24, 35
(see also *historic houses*)
Large scale leisure and tourism
  developments 202, 218–9,
  220–5
Lascaux Caves 190
Leasehold interests (see
  property:interests in)
Leisure
(see also *House of Lords Select*

*Committee on Sport and
  Leisure; large scale leisure
  and tourism developments*)
activities 39, 40, 165, 174, 175,
  205
age 195
as a means of regional
  development 31
centres 1, 51, 168, 176, 201,
  208, 215
complexes 211, 215
definition 3
development 40, 59, 107, 161,
  199, 200, 204, 220
(see also *development*)
ethic 7
facilities 5, 38, 57, 64, 67, 76,
  84, 171, 175, 178
(see also *facility; ice rinks;
  supply of leisure facilities*)
industry 39
market 38, 202
(see also *state intervention in the
  leisure market*)
opportunities 45, 174
planning process 206
policy 21
pools 171, 211, 215
professionals 71
(see also *professionalism*)
property 36, 39, 40–5
provision 36–8, 46, 51, 175–6
and quality of life 17
society 1
Leisure Sport Ltd 38, 64, 200–09
Leisure Studies Association 1
Life cycle cost 82–6, 90, 96, 99
Listed buildings 216
Local
  authorities 11–18, 22, 32, 38,
    45, 51, 54, 58, 61–7, 70–1,
    103–4, 120, 122, 125, 127,
    128, 130, 138, 141–4, 154,
    158, 164–7, 171, 176, 177,
    180, 199, 201, 212–21, 225

(see also *compulsory competitive tendering; House of Lords Select Committee on Sport and Leisure*)
government reorganisation 14, 129, 205
planning authorities 40, 118, 124, 132, 164, 180, 203–8, 210, 213, 220, 222
planning committees 205, 211
plans (see plans:local)
Local Government Act 1972 129
Local Government Act 1988 165, 179
Local Government Planning and Land Act 1980 104
Local Plans Advice Notes 131
Location 2, 35, 105, 201
(see also *national parks:location*)
London Arena 60
London Federation of Ramblers 10
Lotteries 15

Madame Tussaud's 38, 216–221
(see also *Alton Towers; Chessington World of Adventures; Royalty and Empire; Royalty and Railways; Wookey Hole Caves*)
Management
  contract 105
  of property assets 36
Manchester and District Federation of Ramblers 10
Marinas 158–161, 185, 195
(see also *water*)
Market 7, 35, 225
(see also *ideology of the market; land market; leisure market*)
  for development land 135
  economy 16, 164, 172, 176
  failure 32, 36, 45, 168
  (see also *economics of sport and recreation*)
  imperfections 167, 170, 171

  mechanism 67, 116, 157
  price 35, 168
  property (see property market)
  research 62
  segmentation 69, 73, 74
  target 92
Marketing
  process 92
  strategy 74
  techniques 73
Maslow, A. 176
Maslow's Hierarchy of Needs 172–6
Meadowhall shopping centre, Sheffield 61
Merit goods 56, 117, 123, 168–70, 176
Michael Sobell Sports Centre 17
Middle classes 7–9, 13, 17, 144, 163
Militarism 7, 9
Ministry of Agriculture, Fisheries and Food 148, 159, 160
Ministry of Housing and Local Government 127
Mixed
  economy 164
  retail and exhibition centres 195, 218
  retail and hotels 195
  shopping and leisure 19, 57, 108, 111, 195
Municipal parks (see parks)
Museums 19, 58–9, 78

National Countryside Recreation Survey 69
National Economic Development Office 220, 226
National Environmental Protection Act 1969 182, 183
National Farmers' Union 59, 148
National Fitness Council 9
National Motor Museum, Beaulieu 94, 150

National Parks 12, 28, 54, 123–6, 135, 140–43, 157, 225
(see also *access in national parks*)
  authorities 133
  economic efficiency 84
  location 125
National Parks and Access to the Countryside Act 1949 12, 125, 127, 138, 141
National Parks Commission 13, 126, 141, 142
National Playing Fields Association 14, 122, 127
National Rivers Authority 157, 158
National trails 126, 144–5
National Trust 11, 28, 54, 58, 63, 93, 121, 126, 139, 142, 163, 187–92
(see also *Hill, O.*)
National Water Council 155, 156
National Water Sports Centre, Holme Pierrepont 51, 200
Nationalism 9, 16
Nature Conservancy Council 163, 191
Nature reserves 133, 209
Need 2, 65, 67, 71, 165–80, 223
(see also *Maslow's Hierarchy of Needs*)
  affiliation 174
  citizenship 175
  comparative 67, 70
  esteem 174
  expressed 67, 69
  felt 67, 69
  local 214
  normative 67, 69
  physiological 174
  safety 174
  self-actualisation 174
  societal 43, 49
New towns 123, 124
Norfolk Broads 192

Olympic games 9, 171
Open space 54, 120–3, 134–9, 170, 171, 186, 213
(see also *access to open country; urban open space*)
Organisation for Economic Co-operation and Development 46, 183
Outdoor Recreation Resources Review Committee 13
Output 76, 79
(see also *social output of leisure services*)
  final 78, 79
  intended final 79, 100
  intermediate measures 78, 79, 180
  measurement 167
  physical 78, 79
Oxford Farming Conference 148

Park Hall Leisure plc 200, 208
(see also *Somerford Keynes Holiday Village*)
Parks 1, 7, 8, 51, 138, 165–7, 170–6, 194, 201, 225
Participation 177, 179
(see also *planning:public participation; recreation:participation*)
  in leisure activities 54, 69, 175
  in sport and recreation 18, 165, 171, 176
  target rates 180
  trends 69, 177
Peak District National Park 1, 141
Performance
(see also *financial performance*)
  assessment 84, 112, 113
  criteria 168, 180
  evaluation 90, 115, 179
  measurement 112, 113
  measures 39, 179

monitoring 179
Physical
    exercise 180
    fitness 7, 11
Physical Training and Recreation
    Act 1937 12
Picnic sites 13, 143
Place of sport and recreation in
    people's lives 13
Planning
(see also *action areas; enforcement
    notice; general development
    orders; leisure planning
    process; stop notice; use
    classes order*)
    aesthetic considerations 116
    agreements 136
    delays 222
    for leisure 67
    for recreation development 117
    gain 31, 47, 56, 57, 61, 62, 103,
        136, 164, 186, 207
    justification 117
    material considerations 118
    obligations 31
    process 15, 165
    public participation 69, 196
    rural 135
    system 15, 116, 120, 160, 161,
        185, 202, 213, 219, 223, 224,
        225
Planning Advisory Group 127–8
Planning Balance Sheet 99
Planning and Compensation Act
    1991 31, 118, 133, 224
Planning Policy Guidance Notes
    133–4, 225
Plans
    local 118, 128–34
    structure 117–18, 128–34, 143,
        160
Playgrounds 11, 167
Playing fields 1, 11, 51, 165
Pleasure grounds 64
Pollution 153, 192, 197

control 153
Potential
    of recreation and tourism to
        create employment 16
    of sport to ameliorate the
        boredom of unemployment 16
Prices
(see also *market price; role of
    pricing in visitor control*)
    admission 93
    relative 91, 92
    time differentiated 92
Pricing 79, 91, 92
    cost-plus 93
    decisions 93, 94
    differential 92
    discretionary 92
    full-cost 93
    marginal cost 93
    policy 91
    strategy 92
    structure 92
Private
    landowners 13
    (see also *historic houses*)
    sector 54, 65, 83, 93, 168, 179,
        181
Producers' goods 49
(see also *economics of sport and
    recreation*)
Professionalism 10
Promotion of excellence 45
Property 23–6
(see also *leisure property;
    value:property prices*)
    developers 35
    development 41, 42, 47
    (see also *development*)
    easements, covenants and
        mortgages 27
    heterogeneity 36
    interests in 23, 27, 29
    market 35–42
    (see also *market*)
    nature of 37

power 28–32, 223
rental growth 39
rights 12, 26, 27, 33, 34, 141,
    160
taxes 170
Protestant work ethic 8
Provision
    of facilities 11, 14, 22, 62, 87,
        99, 164, 165, 174, 176
    (see also *facilities*)
    motives 56
    of services 165
Public
(see also *planning:public
        participation*)
    access to the countryside (see
        access to the countryside)
    arts policies 171
    footpaths 27
    (see also *rights of way*)
    goods 53, 170–1
    (see also *economics of sport and
        recreation*)
    health legislation 11, 30, 120
    investment in the arts 60
    involvement in statutory
        planning (see planning:public
        participation)
    leisure development 54
    leisure policy 46, 172
    leisure provision 53, 167, 168,
        171–8
    (see also *facility:public sector*)
    liability 30
    open space 117, 121–7, 214
    (see also *development of public
        open space; open space*)
    parks (see parks)
    policy 12, 28, 171
    recreation provision 12, 140,
        201, 224
    sector 15, 19, 20, 38, 45, 47, 56,
        65, 76, 93, 95, 164, 165, 172,
        177, 179, 223, 226
    (see also *ideology of welfare*)

    sports facilities 15, 168, 171,
        211, 220
    transport services 139, 144

Quality
    assurance 113
    management systems 114
    of life 168, 175
Quasi-welfare sector 47
Queen's Valley Reservoir, Jersey
    193

Ramblers' Association 10, 12, 139
Rambling clubs 10
Rank Organisation 38, 57, 75
Recreation 47, 54, 138
(see also *water recreation*)
    activities 7, 8, 10, 49, 71
    centres 75
    concept 3, 7
    development 49, 51, 54, 91,
        103, 138, 185, 195–6, 204,
        224–6
    facilities 1, 7, 12, 13, 49, 72,
        102, 154
    (see also *facility; ice rinks*)
    future role 134
    informal 13, 201, 212, 214
    opportunities 10, 22, 223
    outdoor 4, 138–43, 151–2
    (see also *access*)
    outings 151
    participation 168
    planning 123, 128, 132, 160,
        225, 226
    provision 2, 7, 8, 11, 13, 130,
        157, 201, 223–6
    (see also *public recreation
        provision*)
    use of canals 154–5
    use of gravel pits 151, 199, 202
Redundant barns 47
Regional Councils for Sport and
    Recreation 14, 15, 70, 117,
    155, 178

Regional economic development
    14, 59, 65
Regional Guidance 133
Regional plans for sport and
    recreation 15, 70, 155
Rental growth (see property:rental
    growth)
Rights 45
(see also *access rights*)
    of appeal 119, 124
    of way 28, 62, 123, 138, 142,
        144, 146, 162, 165, 186
    (see also *access; public footpaths*)
Rights of Way Act 1990 162
Risk 103–8
RMC Group plc 38, 64, 202, 221
Royal and Ancient Golf Club 19
Royal Commission on Common
    Land 142
Royal Society for the Protection of
    Birds 63
Royal Yachting Association 152
Royalty and Empire 202, 215–21
(see also *L & R Leisure; Madame
    Tussaud's*)
Royalty and Railways (see Royalty
    and Empire)
Running 165, 175, 178, 180
    tracks 175–6, 215

Sailing 152–4, 199, 202
Sale
    of assets 59
    and leaseback 104
Schools 12, 14, 15, 51, 177
(see also *joint provision*)
Scott, MJ. 12, 123, 141
Scottish Sports Council 85
Seaside resorts 18
(see also *holiday; tourism*)
Season tickets 92
Second homes 31
Secretary of State for the
    Environment 119, 133, 155,
        156, 211, 214, 220, 221

Sherwood Forest 208, 222
Shopping and leisure (see mixed
    shopping and leisure)
Sites of Special Scientific Interest
    191, 220
Skiing facilities 184, 188, 195
Social
    benefits 97
    benefits of tourism 198
    conditioning and control 10
    cost 82, 94
    deprivation 17, 45
    engineering 17, 18
    equity 53
    impact of recreation
        development 226
    output of leisure services 180
    palliatives 17
    policy 46, 144
    value 135–7, 175
    welfare 36, 54, 76, 167, 169, 176
Somerford Keynes Holiday Village
    200, 207
(see also *Granada Group plc;
    Lakewoods Ltd; Park Hall
    Leisure plc*)
South West Water plc 192
Sport 47, 54
    benefits of 169
    ethos 4, 9
    method of social control 10, 16
    and nature conservation 189
    part of welfare state 15
Sport For All 9, 21, 53, 171–9
Sports
(see also *House of Lords Select
    Committee on Sport and
    Leisure; water sports*)
    centres 13, 38, 83–5, 177, 192,
        201
    complexes 39, 195
    facilities 7, 11, 15, 16
    (see also *facility; ice rinks; public
        sports facilities*)
    grounds 11

halls 49, 165, 167, 212, 215, 225
participation 18
stadia 167, 192
Sports Council 13, 14, 16–18, 31,
    83–8, 100, 103, 106, 155,
    162–7, 171–8
(see also *promotion of excellence*)
Squash 19, 40, 73
    courts 211, 215, 225
    and tennis centres 171
Standardised Approach to Sports
    Halls 87, 100, 106
Standards 67, 68
State
(see also *government; public*)
    intervention in the leisure market
        169, 176
    leisure policy 53, 174
    leisure provision 21, 70, 176
Stock market 35
Stonehenge 29, 190
Stop Notice 119
Supply
    of land 36
    of leisure facilities 38
Surveys
(see also *national countryside
    recreation survey*)
    attitudinal 69, 75
    consumer 98
Swimming 153, 175, 178, 180
    pools 1, 11, 20, 49, 51, 85, 165,
        167, 171, 175–7, 211, 215,
        225

Task Force on Sport 9
Taxes
(see also *property taxes*)
    Betterment Levy 61–2, 122, 136
    capital gains tax 128
Tennis 9, 63
Thames Barrier 151
Theatre 7, 8, 38
Theme parks 1, 19, 20, 161, 195,
    200, 202, 207

Thorpe Park 38, 200–7, 219, 222
(see also *Leisure Sport Ltd*)
Timeshare developments 158
Total quality management 113
Tourism 7, 14, 18, 54, 59, 163
(see also *farm tourism; government
    support for tourism; holiday;
    large scale leisure and tourism
    developments; social benefits
    of tourism*)
    destinations 206
    development 30, 60, 132, 161,
        183, 191–4
    economic benefits 31
    economic impacts 133
    employment generation 16, 133
    environmental effects 193
    industry 193, 216, 222, 226
    regions 19
    season 217
    sustainable development 194–5
Tourist informations centres 54
Town and Country Planning Act
    1947 12, 118, 123, 124
Town and Country Planning Act
    1968 128–9
Town and Country Planning Act
    1971 41, 118, 143, 222
Town and Country Planning Act
    1990 118
Town and Country Planning
    (Assessment of Environmental
    Effects) Regulations 1988 184,
    209
Town and Country Planning
    General Regulations 1976 213
Town and Country Planning
    (Minerals) Act 1981 200
Traffic
    congestion 206
    generation 162, 204
Transport 11
(see also *public transport services*)
Transport Act 1962 154
Transport Act 1968 154

Treaty of Rome 182
Trespass 12, 140
  Mass 140

United Nations Conference on the
    Human Environment,
    Stockholm 183
Urban
  aid 16
  commons 11
  containment of urban areas (see
    green belts)
  middle class (see middle classes)
  open spaces 223
Use Classes Order 119
Uthwatt Report 123
Utility 35

Value 34, 35, 175
(see also *development value; land
    value; social value*)
  amenity 191
  ascribed to property 35
  for money 87, 112, 115
  of leisure activity 168, 175–8
  monetary 81
  property prices 192
  surrogates 98
  wildlife 191
Values 172
Visitor centres 54
Visitors
  impact on historic sites 189
  return 206
  role of pricing in visitor control
    94
Voluntary
  clubs 10, 11, 63, 64
  groups 11
  sector 13, 14, 38, 63, 65, 67, 94,
    103

Walkers (Access to the
    Countryside) Bill 1985 145
Walking 50, 152, 160, 165, 177

Warwick Castle 218
Water
(see also *access to reservoirs and
    water courses; Heneage
    Committee; marinas;
    National Rivers Authority;
    National Water Council;
    recreation:use of canals;
    recreation:use of gravel pits*)
  authorities 126, 153–7
  based recreation activities 40,
    151, 152, 192
  bathing in reservoirs 153
  canal network 152–6
  canal users 158
  companies 126, 157, 192
  cruising waterways 155, 158
  flumes 177, 205, 211
  national planning of supply 155
  recreation 154–9, 211
  reservoirs 152–6
  skiing 202
  sports facilities 154, 208
Water Act 1973 155
Water Act 1989 157–8, 192
Water Resources Act 1963 154
Water Space Amenity Commission
    156–7
Welfare
(see also *quasi-welfare sector;
    social welfare*)
  benefits 65, 97
  economics 170
  elements of recreation provision
    223
  needs of society 178, 180
  policy 176
  services 46
  of society 178
  status of leisure provision 21
Welfare State 1, 12, 16, 20, 46,
    124, 167–8, 225
(see also *House of Lords Select
    Committee on sport and
    leisure; sport*)

Welsh Water Authority 156
Wildlife and Countryside Act
     1981 147
Willingness to pay 98
Windsor Castle 216, 218
Wolfenden Committee on Sport
     13, 17, 53
Wookey Hole Caves 218

Work 9
World Leisure and Recreation
     Association 1
World water ski championships
     204, 205

Young Communist League 140
Youth Hostels Association 142